Infrared Observation of Earth's Atmosphere

Series Editor
André Mariotti

Infrared Observation of Earth's Atmosphere

Hervé Herbin
Philippe Dubuisson

ISTE

WILEY

First published 2015 in Great Britain and the United States by ISTE Ltd and John Wiley & Sons, Inc.

ISTE Ltd
27-37 St George's Road
London SW19 4EU
UK

www.iste.co.uk

John Wiley & Sons, Inc.
111 River Street
Hoboken, NJ 07030
USA

www.wiley.com

Library of Congress Control Number: 2015952411

British Library Cataloguing-in-Publication Data
A CIP record for this book is available from the British Library
ISBN 978-1-84821-560-3

Contents

Acknowledgements

The authors would like to thank their colleagues at the Laboratoire d'Optique Atmosphérique (LOA) UMR CNRS 8518 at the Université de Lille 1: Olivier Pujol, Fanny Minvielle and Laurent Couillard-Labonnote, for their continuous advice and generous proofreading as well as Solenne Abraham-Herbin for her contribution to many figures and illustrations featured in this book.

List of Symbols and Units of the Parameters Employed in This Book

Airmass factor:	m
Albedo:	α
Altitude:	z [m]
Asymmetry factor:	g
Azimuth angle:	ϕ
Bidirectional Reflectance Distribution Function (BRDF):	$\rho(s,s_i)$
Brightness temperature:	T_B [K]
Complex index of refraction:	m
Cosine of zenith angle:	$\mu = cos\theta$
Effective radius:	r_e [m]
Emissivity:	ε
Exitance:	M [W.m^{-2}]
Extinction, scattering, absorption coefficient:	k_{ext}, k_{scat}, k_{abs}[m^{-2}.kg^{-1}]
Extinction, scattering, absorption cross sections:	σ_{ext}, σ_{scat}, σ_{abs} [m^{-2}]
Extinction scattering, absorption efficiencies:	Q_{ext}, Q_{scat}, Q_{abs}

Extinction, scattering, absorption

Frequency:	ν [Hz]
Gas density:	ρ [kg.m^{-3}]
Global Warming Potential GWP:	GWP
Height scale:	H [m]
Legendre polynomials coefficient:	β_l
Life time of a species x:	τ_x [s]
Liquid Water Content:	Q_L [kg.m^{-3}]
Molar mass species x:	M_x[kg.mole^{-1}]
Mueller matrix:	$\boldsymbol{S}$
Net flux density:	F^{net}[W.m^{-2}]
Optical thickness:	τ_{ext}, τ_{scat}, τ_{abs}
Particle radius and diameter:	r, D [m]
Planck function:	B_λ [W.m^{-2}.sr^{-1}.µm^{-1}]
Pressure:	p [Pa - hPa]
Radiative efficiency:	RE [W.m^{-2}.ppb^{-1}]
Radiative forcing:	ΔF [W.m^{-2}]
Reflectance:	ρ
Scattering angle:	Θ
Scattering phase function:	$p(\Theta)$
Single scattering albedo:	ϖ_0
Size distribution:	n [m^{-3}]
Size parameter:	x
Solar constant:	F_0 [W.m^{-2}]
Solid angle:	Ω[sr]
Spectral flux density (monochromatic):	F_λ[W.m^{-2}.µm^{-1}]
Spectral intensity (monochromatic):	I_λ[W.sr^{-1}.µm^{-1}]
Spectral radiance (monochromatic):	L_λ[W.m^{-2}.sr^{-1}.µm^{-1}]
Spectral radiative flux (monochromatic):	Φ_λ[W.µm^{-1}]
Spectral source function (monochromatic):	J_λ[W.m^{-2}.sr^{-1}.µm^{-1}]
Surface:	S [m^2]

Temperature:	T [K]
Truncation coefficient:	f
Volume Mixing Ratio (VMR):	VMR [ppmv]
Wave vector and number:	$\boldsymbol{k}, k$ [m^{-1}]
Wavelength:	λ [m]
Wavenumber:	$\tilde{\nu}$ [cm^{-1}]
Zenith angle:	θ

List of Acronyms

3MI:	Multi-viewing Multi-channel Multi-polarization Imaging mission
ACE-FTS:	Atmospheric chemistry experiment Fourier transform spectrometer
AFGL:	Air Force Geophysics Laboratory
AIRS:	Atmospheric infrared sounder
AU:	Astronomical unit
ATOVS:	Advanced TIROS operational vertical sounder
ATMOS:	Atmospheric trace molecules spectroscopy experiment
A-TRAIN:	Afternoon train
ATSR:	Along track scanning radiometers
AVHRR:	Advanced very high resolution radiometer
BOA:	Bottom of atmosphere
BRDF:	Bidirectional reflectance distribution function
BTD:	Brightness temperature difference
CALIOP:	Cloud-aerosol lidar with orthogonal polarization

CALIPSO:	Cloud-aerosol lidar and infrared pathfinder satellite observations
CCD:	Charge-coupled device
CERES:	Cloud and Earth radiant energy system
CkD:	Correlated k-Distribution
CLAES:	Cryogenic limb array etalon spectrometer
CPR:	Cloud profiling radar
DFG:	Deutsche ForschungsGemeinschaft
DOFS:	Degree of freedom for signal
DOM:	Discrete Ordinates Method
ECMWF:	European Centre for Medium-range Weather Forecasts
ERBE:	Earth radiation budget experiment
ESA:	European Space Agency
ESFT:	Exponential sum fitting of transmission
ETHER:	Thematic Expertise Group for Atmospheric Chemistry
FLIP:	Fictive light particle
Flx:	Flux density
GEO:	Geostationary orbit
GISS:	The NASA Goddard Institute for Space Studies
GOES:	Geostationary orbital environmental satellite
GOMOS:	Global ozone monitoring by occultation of stars
GOSAT:	Greenhouse gases observing satellite
GWP:	Global warming potential
HALOE:	Halogen occultation experiment

HIRDLS: High resolution dynamics limb sounder

HSR: High spectral resolution

HWHM: Half width at half maximum

I3RC: Intercomparison of 3D radiation codes

IASI: *Interféromètre Atmosphérique de Sondage Infrarouge* (infrared sensing atmospheric interferometer)

ICAO: International Civil Aviation Organization

ICARE: Interactions cloud aerosol radiation etc.

ICRCCM: Intercomparison of radiation codes in climate models

IFOV: Instantaneous field of view

IIR: Imaging infrared radiometer

ILAS: Improved limb atmospheric spectrometer

ILS: Instrumental line shape

IMG: Interferometric monitor for greenhouse gases

IPA/ICA: Independent pixel/column approximation

IPCC: Intergovernmental Panel on Climate Change

IR: Infrared

IRS: Infrared sounder

ISAMS: Improved stratospheric and mesospheric sounder

ISCCP: International Satellite Cloud Climatology Project

Jac: Jacobian

JAXA: Japan Aerospace Exploration Agency

JPL: Jet Propulsion Laboratory

LBL: Line-by-line algorithm

LEO: Low Earth orbit

LLS: Linear lesat square

LMD: *Laboratoire de Météorologie Dynamique*

LMU: Ludwig Maximilians University, Munich

LOA: *Laboratoire d'Optique Atmosphérique*

LTE: Local thermodynamic equilibrium

LUT: Lookup table

MAP: Maximum a *posteriori*

MAPS: Measurement of air pollution from satellites

MC: Monte Carlo

MERIS: Medium resolution imaging spectrometer

MIPAS: Michelson interferometer for passive atmospheric sounding

MLS: Microwave limb sounder

MO: Matrix operator

MODIS: Moderate resolution imaging spectroradiometer

MOPITT: Measurements of pollution in the troposphere

MSG: Meteosat second generation

MTG: Meteosat third generation

MW: Microwaves

NASA: National Aeronautics and Space Administration

NLLS: Nonlinear least square

NIR: Near-infrared

NOAA: National Oceans and Atmosphere Administration

OCO:	Orbiting carbon observatory
OEM:	Optimal estimation method
OMI:	Ozone monitoring instrument
OSIRIS:	Optical spectrograph and infrared imager system
PARASOL:	Polarization & anisotropy of reflectances for atmospheric sciences coupled with observations from a lidar
POLDER:	Polarization and directionality of the Earth's reflectance
QDIP:	Quantum dot infrared photodetector
QWIP:	Quantum well infrared photodetector
Rad:	Radiance
RMS:	Root mean square
RTE:	Radiative transfer equation
SAGE:	Surface and atmosphere geochemical explorer
SAMS:	Surface atmospheric measurement system
ScaRaB:	Scanner for radiation budget
SCIAMACHY:	Scanning imaging absorption spectrometer for atmospheric chartography
SEVIRI:	Spinning enhanced visible and infrared Imager
SH:	Spherical harmonics
SMOW:	Standard mean ocean water
SNR:	Signal-to-noise ratio
SOS:	Successive orders of scattering
SSM/I:	Special sensor microwave imager

TANSO-FTS:	Thermal and near infrared sensor for carbon observation – Fourier transform spectrometer
TES:	Tropospheric emission spectrometer
TOA:	Top of atmosphere
TRMM:	Tropical rainfall measuring mission
Trs:	Transmission
UCSB:	University of California, Santa Barbara
UW:	University of Wyoming
UV:	Ultraviolet
VIS:	Visible
VMR:	Volume mixing ratio
VPIE:	Vapour pressure isotopologue effect
WIS:	University of Wisconsin
WMO:	World Meteorological Organization

Preface

The aim of this book is not for us to rewrite a new reference book on radiative transfer or atmospheric physics and chemistry; the bibliography of this field is already very rich, such as the reference books we mention all chapters or each chapter. Here, the goal is to provide an overview of spatial infrared observations for studies on the Earth's atmosphere. The theoretical and instrumental bases and the numerical methods are summarized, as well as the main application domains, which are illustrated with the help of some concrete examples.

This book is, therefore, aimed primarily at undergraduate or Master's students in Physics and Physico-Chemistry, who will find it easy to acquire basic knowledge in this field, with the help of many illustrations, tables and summaries. However, we also think that it will be a useful resource for PhD students and researchers in related scientific fields to help better place their work in the context of current and future space missions with the help of many examples and bibliographic references contained in this book. Thus, this provides an initial theoretical overview before thorough knowledge is acquired on a specific topic associated with infrared observations of the atmosphere. For this, we selected a non-exhaustive list of key references in the field, which we present at the end of each chapter in a section called "for further information".

Before you start reading this book, we wish to draw the reader's attention to some recurring definitions throughout this book.

Much space is devoted to high spectral resolution measurements. In general, the spectral resolution of an instrument defines the capacity of its measurement to distinguish between two different wavelengths. Thus, there is no definition in the strict sense of what "high spectral resolution (HSR)" is, since it depends on the spectral variation of the object of study. However, in spectroscopy, it is common to use the term high spectral resolution for any measurement with a resolution in the order of magnitude of the Doppler width of the lines. In a broader sense, and in the case of gaseous absorption analysis by infrared remote sensing, we consider that the term HSR can be applied to all measurements, which helps solve structures of rovibrational molecules (typically $\Delta\tilde{\nu} \leq 1\ cm^{-1}$).

In the field of spectroscopy, the parameter and unit which are commonly used to characterize electromagnetic waves are, respectively, the wavenumber and cm^{-1}. These can sometimes be confusing for those working in the microwave or far infrared range of the spectrum, which often use the frequency: ν [GHz], or for those in the visible-UV domain where one is more likely to use wavelength λ [nm or µm]. Thus, in this book, we often use the wavenumber: $\tilde{\nu}$, which is the number of oscillations of the electromagnetic wave per unit length. This term was introduced for the first time by Rydberg, for quantification of emissions from a hydrogen atom. Subsequently, the work on the radiation–matter interaction focused on theoretical and experimental studies of atoms. The wavenumber then became the "reference" and was often expressed in cm^{-1} to avoid having to deal with many decimal places. Since 1960, molecular spectroscopy has been booming, through the use of Fourier transform spectrometers. However, for these, the wavenumber expressed in cm^{-1} provides a direct mathematical relationship with the optical path difference [cm]. Thus, although a little "unusual", the wavenumber in cm^{-1} is still the preferred unit for characterizing infrared measurements with high spectral resolution.

Finally, in this book, we have deliberately omitted the use of the word photon. Indeed, the term "photon" is frequently mentioned in the field of atmospheric radiative transfer. We often find phrases such as "the direction of the photon", "photon trajectory" and "photon mapping". The latter term is particularly used in the case of Monte Carlo codes used for the radiative transfer in heterogeneous atmospheric environments. This misuse of the term photon has been criticized many times in atmospheric radiation literature. In modern physics, it is in fact well known that it is not possible to associate a position or a precise path to a photon and this term only should be used strictly in the context of quantum electrodynamics. Thus, based on a summary note [PUJ 15], which comments on the misuse of this term, we will use the neologism Fictive LIght Particle (FLIP) as a substitute for the term photon, simple terminology but one that does not contradict the fundamental principles of modern physics.

Hervé HERBIN
Philippe DUBUISSON
September 2015

Introduction

Remote sensing, especially in infrared, has boomed over the past few years. This is the result of constant technical and technological developments, including space missions, which require the quality and reliability of satellite platforms and the measuring instruments they carry to be even greater. This obsession with observing the Earth's atmosphere from space is also related to the recent awareness of air pollution problems such as poorer air quality and climate changes whose consequences are at the heart of issues in current societies. Thus, the aim of this book is to provide an overview of the characteristics and the role of infrared remote sensing for observing the Earth's atmosphere.

Different types of instruments for measuring the composition of the atmosphere have been developed since the early 20th Century. The observation of the atmosphere began with the ground measurements and, in particular, balloon measurements. However, the International Geophysical Year 1957–1958 became a new impetus for aeronomy with the launch of the first artificial satellites. Since then, atmospheric observations from satellite platforms have done nothing but grow, along with their uses in the quantification of climatic variables (greenhouse gases, clouds and aerosols), weather variables (water vapor profiles, temperature, wind and cloud cover) and for monitoring air quality (particulate and gaseous pollution) or atmospheric chemistry (trace gases).

Currently, there are a variety of techniques to measure atmospheric composition, each with its own advantages and limitations. Detectable molecules, microphysical properties of measurable particles or accessible altitude range, for example, vary according to the technical characteristics of the instruments, each offering a limited but complementary view of the atmosphere. However, the use of such measurements require that we understand the physics of radiation and its interaction with matter. Thus, electromagnetic radiation, instrumentation, emission properties, absorption or scattering of atmospheric components, the radiative transfer equation and numerical methods of return of geophysical parameters are all essential concepts for the use of remote sensing covered in this book. This is why the first chapter of this book summarizes the main properties of the emission spectrum of the Sun and Earth, describes the thermodynamic structure and gaseous composition of the Earth's atmosphere and summarizes their role in the radiation budget of the climate system. The following chapter is devoted to satellite platforms and embedded instruments that are most commonly used in infrared remote sensing. The peculiarities of the different types of orbits or viewing-modes will be presented in line with their scientific objectives. Chapters 3 and 4 provide basic information on the key parameters (absorption/emission and scattering/polarization) that describe the interaction between radiation and atmospheric compounds. These allow us to form the radiative transfer equation (RTE) which governs the propagation of light through an absorbing and/or scattering medium, presented in Chapter 4. A panorama of numerical approaches to solving the RTE, and a list of codes, databases and references available in the literature or on the Internet are also available. Chapter 5 explains how, through a procedure known as "inversion", the measurements obtained by remote sensing can be used to obtain quantitative information on the composition of the atmosphere. The mathematical basis of the most common retrieval methods will be described briefly and illustrated in the context of measuring the vertical profile of water vapor. Finally, the book concludes in Chapter 6 with examples of applications of infrared remote sensing for the characterization of gases, clouds and aerosols and their implications in atmospheric or environmental issues related to climate, meteorology or air quality.

The readers will find at the end of the book (in the Appendix) a list of acronyms and key features of a non-exhaustive list of past, present and future infrared instruments for the study of gases and atmospheric aerosols.

Finally, a non-exhaustive list of bibliographic references, but representative of the topics in all the chapters, is also provided.

1

Basic Physics of the Atmosphere and Radiation

Earth's climate is governed, under the influence of solar radiation, by the climate system, which is defined as the set of interactions between the atmosphere, oceans and land surfaces. To measure the variability of the climate system, it is essential to have methods of observation, at both the global and local scale, that are capable of providing long time series of data and can serve as a reference for climate forecasting models. Therefore, remote sensing is an essential tool for the study of the climate system by measuring the electromagnetic radiation emitted or reflected by the atmosphere and by surfaces. In this chapter, we address the main theoretical aspects of electromagnetic radiation and its interaction with components of the climate system to provide a basis for the observation of the Earth's atmosphere using remote sensing. We first present the vertical structure and composition of the atmosphere, emphasizing the gaseous species and particles (aerosols and clouds) that have a significant effect on electromagnetic radiation. Next, we describe the main characteristics of the emission spectra of the Sun and Earth and review the essential concepts of radiometry. Finally, the impact of atmospheric components on radiative forcing of the atmosphere is briefly discussed.

1.1. Structure and composition of Earth's atmosphere

1.1.1. *Vertical structure of the atmosphere*

The atmosphere is a gaseous envelope surrounding our planet. Although there is no clear physical boundary between the Earth's atmosphere and space, the upper limit of the atmosphere generally corresponds to the altitude at which the effects of atmospheric components on solid objects become notable when (re-)entering the atmosphere, which are heated by friction between air molecules. This altitude is typically set at 120 km. In the case of aviation, it is set at 100 km and is known as the Kármán line. The pressure of the gas column forming the atmosphere varies around its mean value. At sea level, this value is equal to $p_a(0) = 1.01325.10^5$ Pa = 1013.25 hPa and is referred to as the standard or the normal. Obviously, depending on local weather conditions, $p_a(0)$ varies and changes between extreme values between approximately 870 hPa (depression) and 1080 hPa (anticyclone). Atmospheric pressure decreases with altitude (Figure 1.1(a)), according to an exponential law, and is reduced by a factor of two around 5500 m altitude. In fact, the gas density ρ decreases with altitude z, due to the gravitational forces. On average, with the vertical movements in the atmosphere usually quite low, we can consider that the atmosphere is in hydrostatic equilibrium. For a portion of air of thickness dz, this equilibrium results from the cancelling out of its weight and pressure forces exerted and is expressed as $dp_a = -\rho_a g dz$, where ρ_a is the air density and g the standard acceleration of gravity is assumed to be uniform. If we consider air to be an ideal gas, we obtain: $\rho_a = M_a p/(R\ T)$, with T temperature, $R \approx 8.3145$ K^{-1}.mol^{-1} the ideal gas constant, $M_a \approx 28.97$ g.mol^{-1} the molar mass of air.

Therefore, for a simple isothermal atmosphere at temperature T, we obtain: $\frac{dp}{p} = -\frac{dz}{H}$, which after integration gives $p_a(z) = p_a(0)\exp\left(-\frac{z}{H}\right)$, with $H = \frac{RT}{gM_a}$, the scale height of the atmosphere which represents the increase in altitude required to divide pressure by a factor of 2.72. For example, for T = 290 K, we obtain $H \approx 8.5$ km. This approximation is useful to carry out a simple analytical calculation.

In reality, the temperature of the atmosphere varies with altitude (Figure 1.1(b)), which induces stratification into layers. In the troposphere, T decreases if z increases; the first minimum is reached at the tropopause (upper limit of the troposphere), whose altitude varies with latitude, season and local conditions. On average, it is found at an altitude of about 8 km in Polar Regions, 12 km at mid-latitudes and 18 km in the tropics. The troposphere, which represents 80 to 90% of the total mass of air ($M_{air} \approx 5.15 \times 10^{18}$ kg), is particularly important because, firstly, it is the layer where most of the weather occurs and on the other hand, it is directly linked to human activities. Above the tropopause is the stratosphere; it extends up to 40–60 km altitude and plays a protective role on living organisms since it contains ozone in large quantities (90% of total atmospheric column reaching a maximum at 25 km) absorbs ultraviolet solar radiation of wavelengths less than 300 nm, which are harmful to living organisms. The mesosphere is found above an altitude of 60 km and the thermosphere beyond 90 km. In the thermosphere, highly energetic radiation from the sun can dissociate molecules of air and ionize atoms. It is in this layer that the aurora borealis occur.

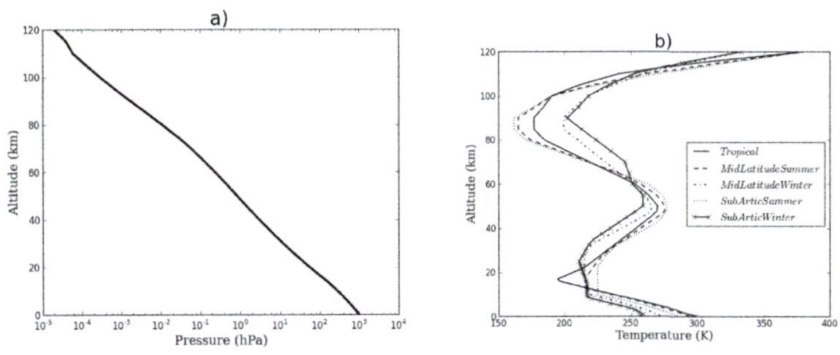

Figure 1.1. *Atmospheric profiles of a) pressure and b) temperature versus altitude. Temperature profiles are presented for standard atmospheres [MCL 72] considered as representative of atmospheric conditions at different latitudes and seasons*

1.1.2. *Atmospheric gases*

Atmospheric air is generally considered to be a mixture of dry air and water vapor, thus emphasizing the special role of H_2O. The "dry" atmosphere is composed mainly of nitrogen (N_2, 78.08%), oxygen (O_2, 20.95%) and argon (Ar, 0.93%); the percentages are expressed as volumes. We can also find other so-called minor gases (less than 0.3% of the atmosphere), whose concentration by volume varies from a few ppmv (parts per million by volume) to tens of pptv (parts per trillion by volume). These are mostly rare gases (neon, helium, krypton, xenon and radon), which are chemically inert and have no influence on atmospheric radiation. Other gases, often in very low concentrations, however, have a significant effect on the absorption of radiation and hence on the greenhouse effect (H_2O, CO_2, O_3 and CH_4) or affect air quality (SO_2, NO_2 and CO). Note that the amount of a gas for a given volume is often expressed as a mixing ratio by volume, denoted in this book as *VMR* [ppmv], defined as the number of molecules of the gas divided by the number of molecules of other constituents of mixture.

Greenhouse gases absorb infrared radiation and significantly affect the radiation balance of the atmosphere. The most absorbent is water vapor, which, contrary to popular belief, is the main greenhouse gas. It represents about 60% of the total natural and anthropogenic contributions, compared with 25% for CO_2. Almost all of the water vapor is in the troposphere and its concentration by volume, which decreases rapidly with altitude, varies on the Earth's surface between 0.5 and 5% of the atmospheric air, depending on the location and season. In addition to water vapor, the main greenhouse gases are: carbon dioxide (CO_2), methane (CH_4), ozone (O_3), nitrous oxide (N_2O) and hydrofluorcarbons (gases consisting of fluorine, chlorine or bromine), such as chlorofluorocarbons (CFC) and hydrofluorocarbons (HCFC), which also have a destructive effect on the ozone layer. Their concentration has been increasing since the 19th Century, mainly because of human activities. Carbon dioxide (CO_2, 0.04%) is the main greenhouse gas (by quantity) produced by human activity. Also, note the presence of reactive gases, carbon derivatives (volatile organic compounds VOC's), nitrogen (NO_x, NH_3) or sulfur (SO_2 and H_2S). Some of these gases contribute to the photochemical formation of ozone. They can also oxidize to form acid (in gas or liquid phase) that contribute to the acid character of the air or precipitation (H_2SO_4, HNO_3, etc).

Some characteristics of the main greenhouse gases are presented in Table 1.1; it shows that the total concentration of gas from anthropogenic origin in the atmospheric column substantially increased during the Industrial Revolution. Note that water vapor is not a gas of anthropogenic origin and we cannot therefore estimate its emissions or effects on climate warming. Moreover, as for ozone, its lifetime in the atmosphere is very low. Figure 1.2 presents the vertical profiles of several atmospheric gases; we can distinguish between gases whose concentration varies with altitude (H_2O and O_3) and those whose concentration remains relatively constant (CO_2, CH_4 and O_2).

Gas	Concentration (ppmv)		Lifetime	Global warming potential (GWP) per 100 years	Radiative forcing in $W.m^{-2}$ (relative contribution)
	Pre-industrial revolution	Current			
Water vapor (H_2O)	constant		day/week	Undefined	Undefined
Carbon dioxide (CO_2)	280	395	100--300 years	1	1.88 (55%)
Methane (CH_4)	0.7	1.8	12 years	28	0.48 (15%)
Nitrous oxide (N_2O)	0.24	0.34	120 years	265	0.16 (5%)
Tropospheric ozone (O_3)	0.24	0.34	hour/day	Undefined	0.35 (10%)
Chlorofluorocarbon (CFC-12)	0	0.0005	100 years	8100	0.17 (5%)
Sulfur hexafluoride (SF_6)	0	0.000008	3200 years	23500	0.0043 (0.1%)

Table 1.1. *Main greenhouse gases in the Earth's atmosphere and their properties. The lifetime corresponds to the activity period of these gases in the atmosphere. Radiative forcing and the global warming potentiall are defined in section 1.5.3*

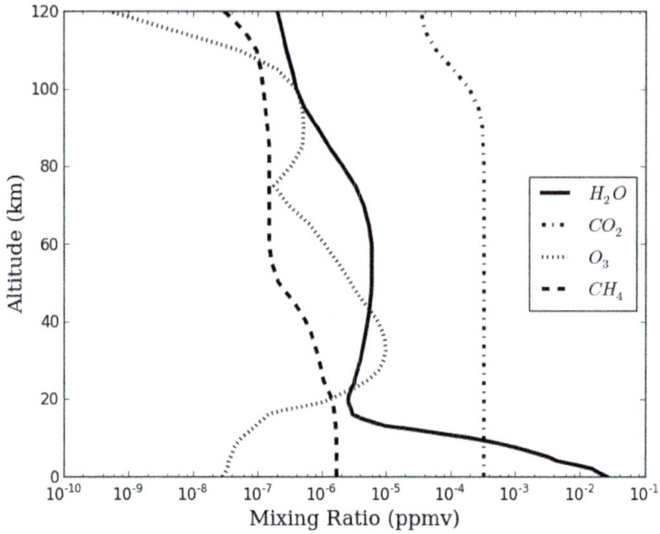

Figure 1.2. *Vertical profiles of the mixing ratios of selected species for a tropical atmosphere [MCL 72]*

1.1.3. *Aerosols and hydrometeors*

The atmosphere contains both liquid and solid suspended matter: aerosols and hydrometeors. Their characteristic size ranges from about 10^{-2} µm for solid particles called fine particulate matter to sizes up to a few mm for larger matter such as hailstones. The term "aerosol" refers to solid or liquid particles suspended in a gas. In general, we exclude solid or liquid aerosols formed mostly of water from this category that are in high concentrations and visible in the sky (that is to say the clouds) because their dominance in our atmosphere is a separate field of study. Hydrometeors are therefore all forms of solid or liquid water present in the atmosphere. We distinguish liquid hydrometeors (rain, drizzle and cloud droplets) from those made of ice (ice crystal, graupel and hail). As the size of aerosols and hydrometeors are generally lower or of the same order of magnitude as the wavelength of atmospheric radiation, these components have a significant effect on solar and infrared radiation. The microphysical and optical characterization of hydrometeors is possible by measuring the infrared radiation emitted and absorbed or scattered by the particles in suspension. However, large hydrometeors are highly absorbent and there is little scattering in the infrared. For remote sensing applications, infrared radiation is then not sensitive to these particles and does not provide usable information on these constituents. Clouds, aerosols and, indirectly, their gaseous precursors therefore have a first-order impact on the radiative balance of climate and they still represent one of the greatest sources of uncertainty when estimating the radiation budget of Earth according to the Intergovernmental Panel on Climate Change (IPCC), [IPC 2013]. The main properties of aerosols and hydrometeors are discussed in more detail in the following sections. Their optical properties will, in turn, be discussed in Chapter 4.

1.2. Atmospheric aerosols

1.2.1. *Overview*

Atmospheric aerosols are highly variable in form, quantity, chemical and mineralogical composition and life-time. We can distinguish between naturally occurring aerosols (volcanic, desert and sea) from those of anthropogenic origin. Aerosols are also often classified according to whether they are primary or secondary in nature, i.e. whether they are emitted directly into the atmosphere or formed by chemical and photochemical reactions from a gaseous precursor in the atmosphere, respectively. They can also be grouped by internal mixing (chemical species or different minerals in the same particle) or external mixing (different types of particles in the same aerosol plume, each particle of the population being composed of a single chemical compound). Chapter 5 of the IPCC report (2001) provides a classification by type of aerosol source: desert dust; sea salt; industrial dust; carbonaceous aerosols; primary biogenic aerosols; sulfate aerosols; nitrates; and volcanic aerosols.

The chemical composition of aerosols reflects that of the source. One can for example distinguish between aerosols sources whether desert dust, carbonaceous aerosols (derived from biomass burning) or industrial dust. The variety of sources, both natural and anthropogenic, combined with the variety of microphysical properties of aerosols and the dynamic physical and chemical processes involved in their life cycle explain their high spatial and temporal variability. Some regions are sources of recurrent aerosols, due to their geological, climate or industrial characteristics. The main sources of mineral dust are the large deserts (Sahara, Gobi deserts of the Middle East and Australia). The main source areas of carbonaceous aerosols are Africa (north and south of the Equator), South America, Indonesia, Australia, the vast expanses of Siberia. These are regions where biomass fires regularly occur, usually during the dry season in each hemisphere. On the other hand, the increase in population and the strong economic development of our societies favor the emission of anthropogenic aerosols in urban areas because of the combustion phenomena necessary for energy production (domestic heating, industry, transport) or from rural agriculture. Major sources are therefore situated near-industrial centers and large urban concentrations such as northwestern Europe, North America, India and Southeast Asia. Finally, volcanic eruptions are also a source of aerosol emissions, which are more localized in time and space, but can have a significant influence at the local level and in some cases, on a more global scale. The eruption of the Icelandic volcano Eyjafjallajökull (Iceland), which paralyzed European aviation in spring 2010, is a good example of the potential impact these aerosols can have on a global scale. Table 1.2 summarizes the main types of primary aerosols and precursors of secondary aerosols, and their emissions. Note, however, that the effect of aerosols on climate does not only depend on emissions, but also on their characteristics (microphysics and chemistry).

Due to erosion and atmospheric circulation, aerosols go up and are transported in the atmosphere. As the size and therefore the mass of aerosols can be very small (micrometre), aerosols can be transported over long distances. For example, desert dust storms occurring in North Africa may be carried to the other side of the Atlantic in only a few days, at altitudes of 3 to 5 km. In more extreme cases, such as volcanic eruptions, particles can be released into the stratosphere and may reside there for several years; vertical exchanges are slower in this atmospheric layer than horizontal ones. During their transport, aerosols undergo transformations, mixtures, or aging, that modify their physicochemical properties. Aerosols are then deposited on the ground by wet or dry deposition. The latter is the deposition of aerosols on the ground by sedimentation (gravity), impaction or Brownian diffusion. Wet deposition occurs through the prior integration of aerosols into liquid droplets or ice particles (as condensation nuclei) within clouds, or they are washed out during rain events.

Type of aerosol	**Emissions (per year)**
Natural primary aerosols	
Desert dust	1000–3000 Tg
Marine salts	1000–6000 Tg
Aerosols from biomass burning	20–35 Tg
Primary biogenic aerosols	1000 Tg
Bacteria	40–1800 Gg
Spores	28 Tg
Precursors of natural secondary aerosols	
Dimethylsulfide (DMS)	20–40 Tg S
Volcanic SO_2	6–20 Tg S
Terpenes	40–400 Tg
Primary anthropogenic aerosols	
Industrial dust	40–130 Tg
Aerosols from biomass burning	50–90 Tg
Soot (fossil fuels)	6–10 Tg
Organic Carbon (fossil fuels)	20–30 Tg
Precursors of secondary anthropogenic aerosols	
SO_2	70–90 Tg S
Volatile Organic Compounds (COV)	100–560 Tg C
NH_3	20–50 Tg N

Table 1.2. *Approximate emissions of the main types of primary aerosols and precursors of secondary aerosols. Tg = 10^{12} g = 1 million tons; Gg = 10^9 g = 1 thousand tons; C = carbon, S = sulfur, N = nitrogen. Source: [BOU 15]*

1.2.2. *Microphysical properties of aerosols*

Aerosol properties (size, shape and chemical composition) are varied and complex. Aerosols encountered in the natural environment are composed of a large number of particles of different sizes. In the atmosphere, aerosols are present in three modes (Figure 1.3):

1) ultra fine particles (nucleation and Aitken modes): This mode is for particles from the gas-particle conversion whose size ranges from 10^{-3} to 10^{-1}µm. As the size of these particles is significantly less than the wavelength of the atmospheric radiation, their radiative effect will be low. They will, however, have significant impact on human health;

2) accumulation: This corresponds to particles several tenths of micrometres in size. They are formed by the coagulation of finer particles;

3) coarse particulate matter: These are particles with sizes of only a few µm, emitted by surface mechanical processes such as sea salt and desert dust.

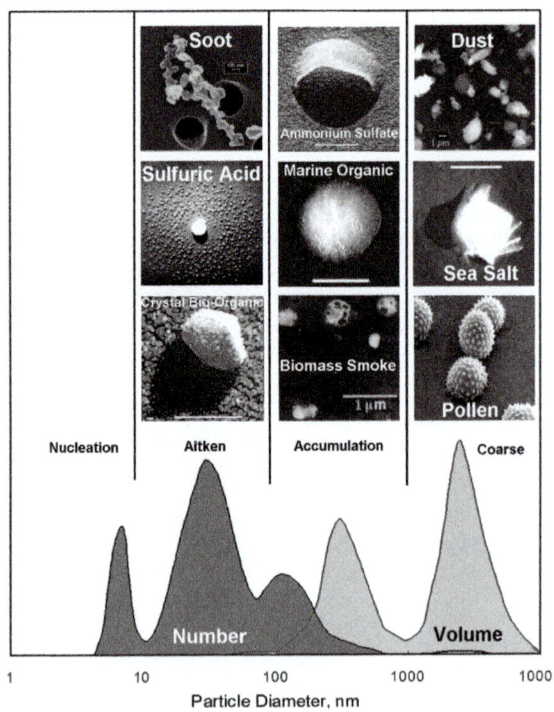

Figure 1.3. *Examples of different types of aerosol and their size distribution. Source: [HEI 03]*

According to the particle size, we can also distinguish between the different classes of aerosols as a function of their particulate matter (PM):

1) suspended particles with a diameter greater than 10 micrometers collect in the upper airway (nose and mouth);

2) PM_{10}: suspended particles with a diameter less than 10 micrometers. These particles, known as respirable, can penetrate the lungs;

3) $PM_{2.5}$: suspended particles with a diameter less than 2.5 micrometers. These fine particles penetrate the pulmonary alveoli;

4) $PM_{1.0}$: suspended particles with a diameter less than 1.0 micrometer. These very fine particles can pass through the alveolar-capillary barrier;

5) $PM_{0.1}$: suspended particles with a diameter less than 0.1 micrometer are called ultrafine or nanoparticles.

These are not single sizes but instead a distribution of the number n of particles according to their size. Experimental observations show that the aerosol size is distributed fairly, according to a lognormal function. This size distribution gives the number of aerosols per unit volume $n(r)$, between the radii r and $r + dr$ and is written as:

$$n(r) = \frac{n_0}{\sqrt{2\pi}\, ln\,(\sigma)\, r} \exp\left[-\frac{\left(ln\, r - ln\, r_m\right)^2}{2 ln^2 \sigma} \right], \qquad [1.1]$$

where r_m the mean radius of the distribution, $ln\ \sigma$ the standard deviation that quantifies the width of the distribution and n_0 the number of particles per unit volume.

Aerosol size is, however, not clearly defined as the particles are not systematically spherical and their shape varies depending on their origin and composition. The shape of aerosols can be seen under the electron microscope and a wide variety and complexity of forms have been observed (Figure 1.3). The spherical approximation is realistic for droplets, but is far from reality for desert dust for example. The difficulties encountered when describing the shape of aerosols include the lack of data and representativeness of a theoretical shape with respect to natural aerosols. A good compromise between the spherical approximation and a more realistic description of aerosol shape is the use of spheroids, which allows for calculations of optical properties for a range of aspherical particles.

If the shape of the particle is complex, the length of the longest axis can be taken as the size. Another widely used approach is to define an effective size, which reflects the size dependence of scattering and absorption processes during the propagation of radiation. We therefore defined the effective radius r_e as:

$$r_e = \frac{\int_0^{\infty} r^3 n(r)dr}{\int_0^{\infty} r^2 n(r)dr}. \qquad [1.2]$$

The interest in the effective radius is due to the fact that it can be set for any particle, regardless of its form, and is the most appropriate variable when studying the aerosol-radiation interaction. For alognormal distribution, we obtain the relation $r_e = r_m \exp\left[\frac{5}{2} Ln^2\sigma\right]$. These size parameters (r_m, $Ln\sigma$, r_e) are another advantage of defining the different distributions using common parameters. Finally, we have the size parameter (or size factor) $x = 2\pi r/\lambda$, where r is the radius of the particle and λ wavelength (see section 4.1.1). The mineralogical and/or chemical composition of atmospheric aerosols is also highly variable. When the aerosol interacts with radiation, it is the refractive index that provides information about the composition of the aerosol and allows us to characterize its optical properties (see Chapter 4). Unfortunately, measurements of the refractive index in the infrared are limited and tricky. They must however be understood in order to link the microphysical and optical properties of aerosols.

1.3. Clouds

1.3.1. *Definitions and classification*

Depending on the temperature of the environment, clouds occur as a cluster of water droplets or ice crystals suspended in the atmosphere. The diameter of the spherical droplets of liquid water is limited to approximately 100 µm, but small values (between 10 and 40 µm) predominate. Ice crystals, with a more complex geometry, are typically between a few micrometres and 2 mm in size. Clouds play an essential role in the energy and water cycles, and therefore, in the energy balance of the atmosphere. The processes of evaporation and condensation are in fact, via the clouds, where a considerable amount of energy is transferred on a local or global scale from the tropics to higher latitudes. For example, the total annual precipitation across the globe is an average height of about 90 cm. The energy required to evaporate this amount of water corresponds to more than a quarter of the solar energy received by the Earth each year. This energy is then returned to the atmosphere during the process of condensation. Moreover, clouds have a significant impact on solar and telluric radiation and even more so when they are widespread and persistent. On average, they permanently cover about 70% of the Earth's surface. They therefore play an important role in remote sensing since 90% of space measurements acquired globally are estimated to be affected by clouds. Space observations have allowed us to estimate that low or medium altitude clouds, composed of liquid water droplets, represent about 55% of the total cloud cover, with the rest corresponding to clouds at high altitude, mostly composed of ice crystals. While the former reflect more solar radiation, essentially causing cooling, the latter, relatively transparent to sunlight, mainly contribute to warming of the atmosphere by trapping some of the infrared radiation emitted by the surface and the lower layers of the atmosphere.

Clouds are classified according to their altitude, their appearance (texture) and vertical development (Table 1.3). This classification (defined by the WMO-World Meteorological Organization) is based on that defined by the British scientist Luke Howard (1803) in the 19th Century. It uses a Latin terminology whose roots provide information on the classification of components: Cirro- (filament), Strato- (layer-elongated), Cumulo- (heap), Alto- (medium), Nimbo- (rain). There are 10 types of clouds, classified into four categories depending on their altitude: the clouds of the upper, middle and lower layers, as well as those with significant vertical development, which occupy several floors. The cloud base strongly depends on the temperature of the atmosphere, and therefore the latitude. Clouds are completely composed of ice below a temperature of -38°C; between 0 and -38°C, they may consist of a mixture of ice and liquid water whose proportion is difficult to predict.

Clouds are therefore key elements in weather forecasting in the short term and in climate change in the long term. However, their variable and complex spatiotemporal structure makes it difficult to accurately model their impact. A poor estimate of the structure of cloud cover at low altitude can cause the surface temperature to be out by several degrees when forecasting a few days ahead. For technical reasons, representing clouds in models is simplified relative to reality: the grid used for numerical weather forecasts (vertical and horizontal resolution of the models) are approximately several kilometers to tens of kilometers in size and are much larger than cloud processes (convection, condensation of water vapor, evaporation of liquid water, collisions of droplets, etc.), which are also influential in the formation and evolution of clouds. There is still significant uncertainty with regard to the microphysical parameterization of clouds in models; the three phases of water may coexist depending on the atmospheric temperature.

The cloud dataset most commonly used for climate studies is the International Satellite Cloud Climatology Project (ISCCP) [ROS 91]. Originally intended to provide a climatology of clouds to study the processes related to clouds on synoptic scales, the ISCCP uses measurements from meteorological satellites to determine cloud cover, classify clouds and determine other properties (optical thickness, top pressure, water content, droplet size, etc.) with a spatial resolution of 280 km, spanning the globe with a time step of three hours.

1.3.2. *Formation*

A cloud is formed by condensation or deposition of water vapor, when the latter has exceeded its saturation point at the given pressure. We refer to the dew point in the case of liquefaction and the frost point for solidification. The formation of a cloud also requires the presence of aerosols that act as condensation nuclei on which cloud droplets will form. There are several ways clouds can form:

1) Formation by adiabatic expansion (without energy from heat exchange): the air near the surface and the water vapor it contains, heated by the sun, rise and are cooled by adiabatic expansion (decrease in atmospheric pressure accompanied by a decrease in temperature), which often occurs via thermal convection. Convection clouds form more easily as there is cold air at altitude (unstable atmosphere). The base of such clouds is horizontal (saturation level) and the cloud top changes according to temperature, up to the tropopause in some cases (cumulonimbus).

2) Frontal lifting: clouds can form along fronts, defined as the interface between two air masses of different meteorological parameters (temperature, pressure and humidity). A mass of warm air moving horizontally may be lifted at altitude, at the front, by the mass of previous denser cold air (warm front). The hot air slowly rises as it surpasses the cold air and clouds tend to form along the front, for example cirrus at altitude and stratus closer to the ground. However, a mass of cold air moving horizontally (cold front) will raise the mass of previous warm more rapidly, forming cumulonimbus at the front of the cold front and cirrus at altitude after the front passes.

3) Orographic lifting: the relief forces the air mass to rise sharply on the windward side. Depending on the moisture content of the air, a cloud may form on windward side and precipitate. It then dissipates on the leeward side.

4) Cloud base cooling: other thermodynamic transformations help saturate a moist air mass and form mists, which, unlike clouds, are in contact with the ground. They may form by isobaric cooling of low atmospheric layers, or during night cooling in clear air (radiation fog)

or when moist air passes over a cool surface (advection fog). There is also a type of fog that forms when the air mass comes in contact with a body of liquid water (evaporation fog). Saturation is reached in the latter case with water vapor from the mass of liquid water. Finally, the horizontal mixing of two air masses can cause the formation of a dense fog (mixing fog).

Category	Name	Base altitude (thickness) in km	Base temperature	Characteristics
High clouds	*Cirrus* (Ci)	6–12 (0.2–1.8)	-20 to -60°C	Non-precipitating clouds in the form of white filaments, made of ice crystals.
	Cirrocumulus (Cc)			Thin layer or sheet of ice crystals or sometimes supercooled water, non-precipitating, composed of small elements in the form of granules.
	Cirrostratus (Cs)			Cloudy transparent veil composed of ice crystals, non-precipitating, fibrous, often resulting in halo phenomena.
Mid clouds	*Altocumulus* (Ac)	2–6 (1.5–2)	+10 to -30°C	Gray or white cloud sheets, partly fibrous, composed of liquid water sometimes with ice, often causing precipitation.
	Altostratus (As)			Cloud layer or sheet, fibrous or uniform often causing precipitation, composed of liquid water and ice, do not produce halo phenomenon.
Low clouds	*Nimbostratus* (Ns)	0–3 (3)	+10 to -15°C	Layer of dark and greyish clouds, accompanied by heavy rain or snowfall that usually reach the surface.
	Stratocumulus (Sc)	0.3–2 (0.6)	+15 to -5°C	Gray or whitish sheets, composed of dark round masses, non-fibrous, often causing precipitation.
	Stratus (St)	0 1.2 (0.3)	+20 to -5°C	Grey cloud layer, uniform, resulting in drizzle or snow grains, causing mist if the base touches the ground.

Vertical development	*Cumulus* (Cu)	0.3–2 (1)	+15 to -5°C	Isolated cloud, white, dense with well-defined contours, tower-shaped or like "cotton wool" which can sometimes cause showers.
	Cumulonimbus (Cb)	0.6–2 (9)		Very dense cloud with strong vertical development in the form of tower or anvil at the top, causing heavy rain, hail, and thunderstorms.

Table 1.3. *Classification of clouds according to the World Meteorology Organization (WMO)*

1.3.3. *Microphysical properties*

The formation of microscopic droplets from pure water vapor occurs via a homogeneous nucleation mechanism, which forms a nucleus less than one micrometer in size. However, this process requires supersaturations, which are very difficult to achieve in the troposphere. Aerosols, which act as condensation nuclei, allow water vapor to liquefy and then form cloud droplets, by a process called heterogeneous nucleation. The growth of the droplets depends on the size and number of these condensation nuclei. Coalescence, the amalgamation of several droplets by collision to form a larger one, then allows the droplets to reach a sufficient diameter to eventually form raindrops with sizes between 200 μm and 3 mm. For liquid droplets, the radius distribution is rather a modified gamma distribution [DEI 69]:

$$n(r) = ar^{\alpha} \exp\left[-\frac{\alpha}{\gamma}\left(\frac{r}{r_c}\right)^{\gamma}\right], \qquad [1.3]$$

where $n(r)$ is the number of droplets per unit volume with radius r. The parameter a represents the total number of drops per volume. The shape of the distribution is given by the empirical parameters α, γ and r_c and controls the fraction of small or large particles (Table 1.4). Figure 1.4 shows the size distribution for several types of clouds.

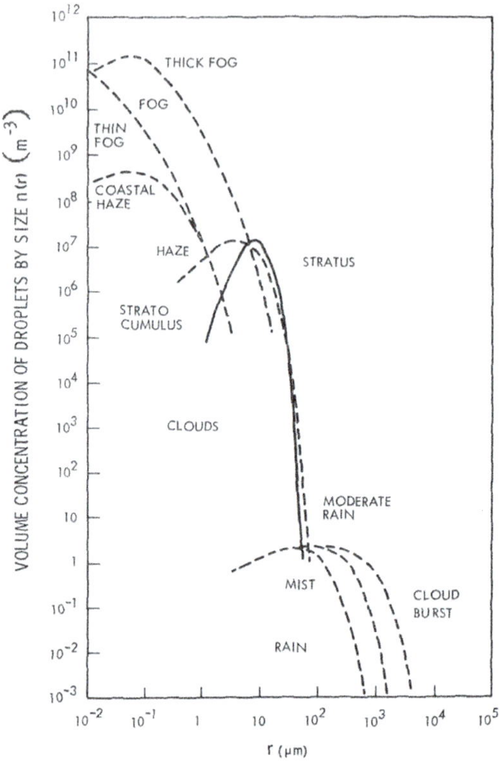

Figure 1.4. *Size distribution functions n(r) for water droplet clouds (adapted from [HUN 71])*

Cloud type	a	α	γ	r_c (μm)
Stratocumulus Sc1	0.2823	5	1.19	5.33
Stratocumulus Sc2	0.19779	2	2.46	10.19
Stratus St1	0.97923	5	1.05	4.70
Stratus St2	0.38180	3	1.3	6.75
Nimbostratus Nb1	0.080606	5	1.24	6.41
Nimbostratus Nb2	1.0969	1	2.41	9.67
Cumulus	0.5481	4	1.0	6.0
Precipitating cloud	$4.97\ 10^{-8}$	2	0.5	70.0

Table 1.4. *Parameters of the modified gamma distribution defined by [DEI 69] for different types of clouds*

In the case of precipitation, the largest drops follow a gamma distribution [ULB 83]:

$$n(D) = n_0 D^{\mu} \exp(-\lambda D), \qquad [1.4]$$

as a function of the diameter D of raindrops, with n_0, μ and λ empirical scaling parameters, of shape and slope, respectively. From the size distribution $n(D)$ and density of liquid water ρ_L [kg.m^{-3}], it is easy to define the liquid water content Q_L [kg.m^{-3}]: $Q_L = \frac{\pi}{6}\rho_L \int_{D_{\min}}^{D_{\max}} D^3 n(D) dD$.

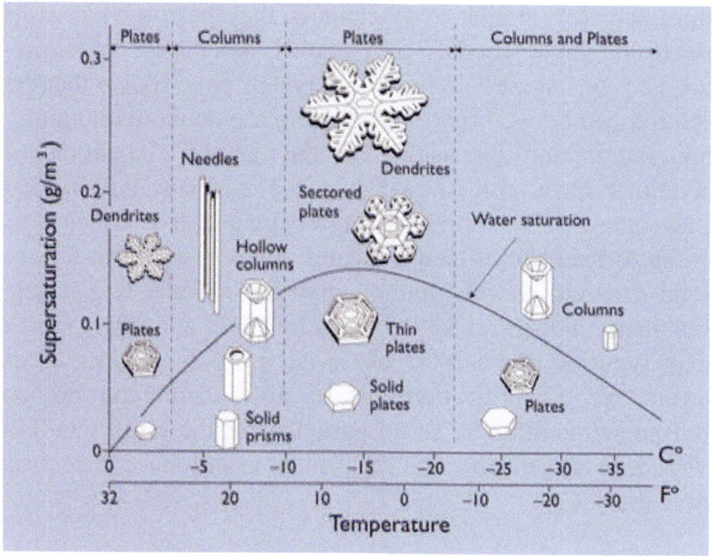

Figure 1.5. *Effects of temperature and humidity on crystal formation. Image credit: Kenneth Libbrecht*

Crystal formation and growth mechanisms are much more complex. In theory, at atmospheric pressure, crystals can form at temperatures below 0°C, either by deposition of water vapor (transition from gaseous to solid state) or by freezing of liquid water droplets (transition from liquid to solid). This homogeneous nucleation process involves only water, without the presence of any another component. Again, homogeneous nucleation is inefficient, especially for deposition, and becomes effective only when droplets freeze at temperatures below about -38°C. Aerosols, also known as ice nuclei, are therefore essential for heterogeneous nucleation. This process also requires a significant degree of supersaturation before deposition can begin. Therefore some drops may remain in the liquid form, in a supercooled state, until temperatures of about -38°C are reached. Then, a supercooled water droplet freezes immediately if it comes into contact with a crystal. Crystals that are not spherical in shape, follow the growth axes, according to the crystallographic properties of the ice, which vary depending on the temperature and degree of supersaturation. Despite the hexagonal crystallographic structure, which is similar in all crystals, they have a wide variety of forms, from single crystals to complex assemblies or aggregates (Figure 1.5). *In situ* measurements of crystals have shown that the size distribution can be well represented using a gamma distribution, with D the maximum dimension of the crystal (equation [1.4]). Observations have also shown that the concentration of hail or graupels (snow in the form of white, opaque ice particles) varies according to the cube of their diameter. Finally, note that for crystals or cloud droplets, the distribution may be bimodal (e.g. using two gamma distributions) to take into account small and large particles or different types of particles, as observed from *in situ* measurements. However, for simplification, it is not uncommon to use a monodisperse distribution, so all particles are the same size. Finally, note that, as with aerosols, it is convenient to introduce the concept of effective diameter.

1.4. Radiation in Earth's atmosphere

1.4.1. *Electromagnetic radiation*

Solar energy reaches our planet in the form of electromagnetic waves. Under the effect of this radiation, the ground and the atmosphere warm up and in turn emit energy in the form of infrared radiation. Radiative exchanges are therefore paramount for the study of Earth's climate system. In addition, detecting distant objects using electromagnetic waves forms the basis of modern remote sensing. An electromagnetic wave is defined by its frequency ν [s^{-1} or Hz] = $1/T$, with T the period [s], and its energy E is equal to $h\nu$, where h is Planck's constant. In addition, the wave polarization state is used to provide information about the orientation of the associated electric and magnetic fields (see Chapter 4). Depending on the nature of the medium through which the electromagnetic wave propagates, we can define wavelength λ, linked to the frequency by the relation $\lambda = v / \nu$, with $v = c / n$ the wave propagation speed, n refractive index of the medium c the speed of light in a vacuum (for which $n = 1$). Short wavelengths refer to the visible spectrum (approximately 0.2 to 4 µm), whereas the thermal infrared region, terrestrial radiation, corresponds to long wavelengths (approximately 4 to 100 µm). Figure 1.6 shows the different regions of the electromagnetic spectrum. Wavelength is usually expressed in nm or µm in the visible near infrared regions. In the infrared spectrum, it is common to use the wavenumber $\tilde{\nu}$, in cm^{-1}. Wavelength and wave number are linked by the formula $\tilde{\nu} = 1/\lambda$, with the following conversion relations: $\tilde{\nu}$ [cm^{-1}] = $10^4/\lambda$[µm] and $\Delta\tilde{\nu}/\tilde{\nu} = \Delta\lambda/\lambda$, with $\Delta\lambda$ and $\Delta\tilde{\nu}$ the spectral steps.

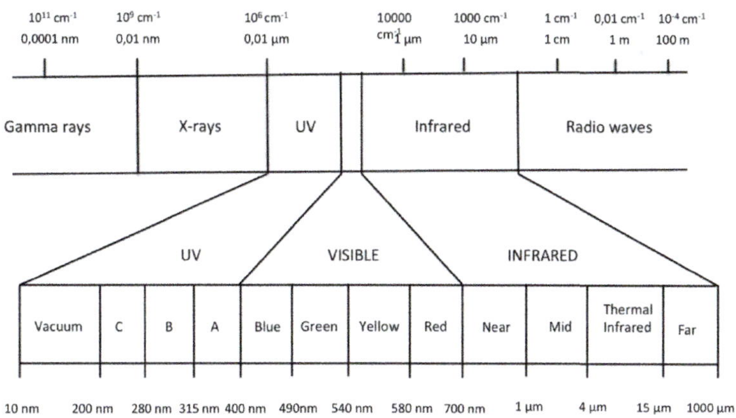

Figure 1.6. *Different domains of the electromagnetic spectrum*

1.4.2. *The foundations of radiometry*

Let us consider an elementary quantity of spectral radiant energy dQ_λ passing through a surface dS, during a time interval dt and in a spectral interval [λ,λ+$d\lambda$] (Figure 1.7). The spectral radiant flux is defined as $\Phi_\lambda = dQ_\lambda/dt$ [W.µm^{-1}]; spectral radiation flux density F_λ is therefore defined as $d\Phi_\lambda/dS$ [W.m^{-2}.µm^{-1}]. Note that the flux density F_λ is often referred to as radiant flux, which is incorrect. In practice, the term irradiance is usually used, instead of flux density, when radiation is incident, and exitance if the radiation is emitted by a surface. The directionality of the emitted radiation by a source is defined by spectral intensity I_λ, [W.sr^{-1}.µm^{-1}], which is the flux per solid angle unit $d\Omega$ [sr]:

$$I_\lambda = \frac{d\Phi_\lambda}{d\Omega} = \int_S L_\lambda \cos\theta dS \text{ and } L_\lambda = \frac{\partial^2\Phi_\lambda}{\partial\Omega\partial S\cos\theta}, \quad [1.5]$$

with L_λ the spectral radiance, defined as the energy per unit surface, per solid angle unit, per wavelength and per unit time [W.m^{-2}.sr^{-1}. µm^{-1}]. Radiance can also be expressed per unit of wavelength [W.m^{-2}.(cm^{-1})$^{-1}$.sr], with the conversion relation: $L_{\tilde{v}} d\tilde{v} = L_\lambda d\lambda$. Of course, we then obtain the flux density [W.m^{-2}] or the radiance [W.m^{-2}.sr^{-1}] in a given spectral band [λ_1, λ_2] by summing up all the wavelengths between λ_1 and λ_2, such that $F = \int_{\lambda_1}^{\lambda_2} F_\lambda d\lambda$.

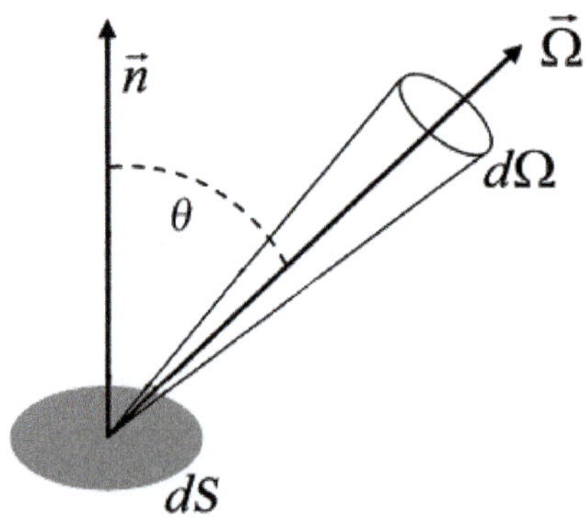

Figure 1.7. *Geometry used to define the radiation passing through a surface dS*

To calculate the radiation budget of a section of atmosphere of thickness *dz*, it is useful to distinguish the incoming and outgoing radiation, which defines the upward hemispheric spectral flux density (upper hemisphere) and downward hemispheric spectral flux density (lower hemisphere), both defined as positive at a specific atmospheric level. This concept helps define the net spectral flux density F_λ^{net}, which is used in the calculation of radiative forcing (see section 1.5.3):

$$F_\lambda^{net} = F_\lambda^+ - F_\lambda^- = \int_{2\pi} L_\lambda^+ \cos\theta d\Omega - \int_{2\pi} L_\lambda^- \cos\theta d\Omega = \int_{4\pi} L_\lambda \cos\theta d\Omega, \quad [1.6]$$

with L_λ^+ and L_λ^- the radiances corresponding to the upper and lower hemisphere respectively, the indices + and – representing upward and downward radiation. The equation is also written in spherical coordinates such as:

$$F_\lambda = \int_0^{2\pi} \int_0^{\pi} L_\lambda(\theta,\phi) \cos\theta \sin\theta d\theta d\phi, \quad [1.7]$$

with θ and ϕ being the zenith and azimuth angles, respectively. We often use the notation. Note that the notation cos $\theta = \mu$ is commonly used.

1.4.3. *Solar and terrestrial radiation*

The spectral radiant energy $L_\lambda(T)$, also denoted as $B_\lambda(T)$, emitted by a black body follows Planck's law, which defines the distribution of the radiance of the thermal radiation of the black body as a function of its thermodynamic temperature and wavelength, assuming thermodynamic equilibrium between the black body and radiation:

$$L_\lambda(T) = \frac{2hc^2}{\lambda^5 \left[\exp\left(\frac{hc}{\lambda kT} \right) - 1 \right]} \quad \left[W.m^{-2}.sr^{-1}.m^{-1} \right], \quad [1.8]$$

with k being the Boltzmann constant. A black body is therefore an ideal environment in which the transmitted electromagnetic spectrum depends only on the temperature, according to Planck's law, absorbing all radiant energy it receives, without scattering or transmitting it. The radiation emitted by a black body is not polarized and is isotropic. For example Figure 1.8(a) shows $L_\lambda(T)$ for temperatures close to those found at the center of a flame, on the surface of the Earth or Sun. Moreover, the brightness temperature (or apparent radiation temperature T_B) of a given surface is defined as the temperature of a black body outputting the same radiance as that measured for this surface at a given frequency.

The total radiation intensity emitted by a black body, integrated in wavelength and over a demi hemisphere, is called the energy exitance M. It can also be calculated by integrating Planck's law, from the total integrated radiance $L(T)$:

$$M(T) = \pi\, L(T) = \pi \int_0^\infty L_\lambda(T)\, d\lambda\,, \qquad [1.9]$$

where the factor π reflects the isotropic nature of radiation (Lambert's law). This relation can be expressed differently by changing the variable $x = (hc)/(k\lambda T)$:

$$M(T) = \frac{2\pi k^4 T^4}{h^3 c^2} \int_0^\infty \frac{x^3}{\left(\exp[x]-1\right)} dx = \frac{2\pi k^4 T^4}{h^3 c^2} \frac{\pi^4}{15} = \sigma T^4\,, \qquad [1.10]$$

where $\sigma = 5.67\ 10^{-8}$ W.m^{-2}.K^{-4} is the Stefan-Boltzmann constant. This expression is known as the Stefan-Boltzmann law. Moreover, the derivative $dL_\lambda(T)/d\lambda$ is cancelled out for a wavelength λ_m corresponding to a maximum of L_λ. We find that $\lambda_m\ T \approx 2.897\ 10^{-3}$.m.K, a relation known as Wien's displacement law. So, for temperatures T_T = 300 K or T_S = 5800 K, which correspond to the surface of the Earth and Sun, respectively, we find a wavelength λ_m equal to 9.6 µm and 0.5 µm, respectively. The spatial distributions of solar and terrestrial radiation are described in the following sections.

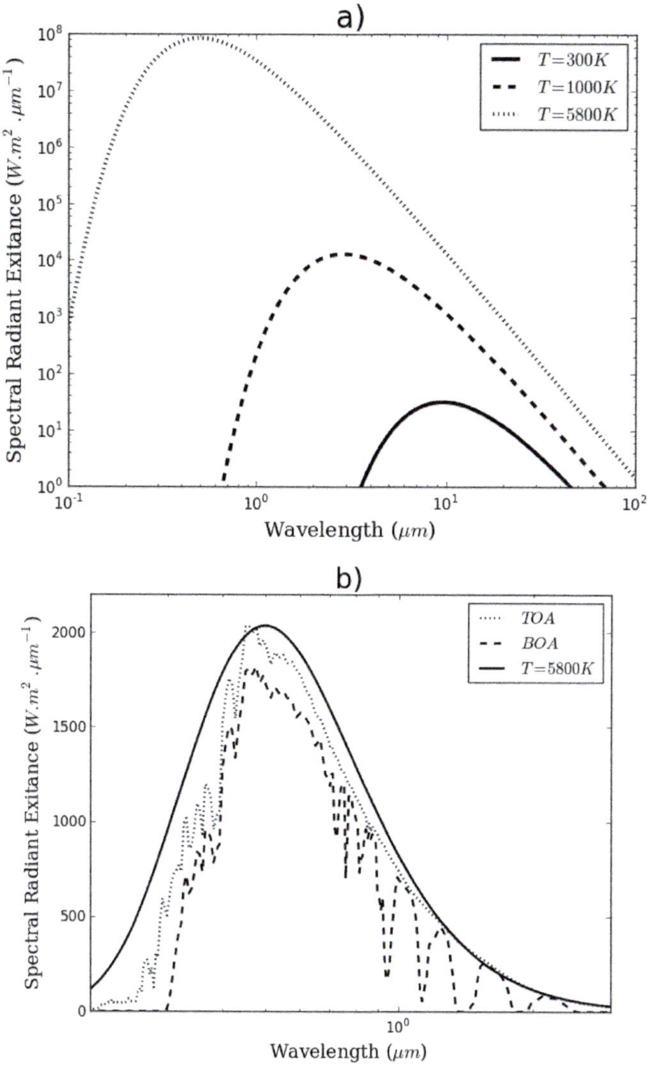

Figure 1.8. *a) Black body emission curves for temperatures close to those found at the surface of the Sun (5800 K), a flame (1000 K) and the Earth (300 K); b) solar irradiance spectrum measured at the top of the atmosphere (TOA) and modeled at sea-level (BOA) in the absence of clouds or aerosols. Planck curve of a black body at 5800 K is also plotted as a solid line*

The Sun radiates like a black body. Most of this radiation comes from the photosphere, the upper gaseous layer of the Sun with a thickness of a few hundred kilometers and a temperature of approximately 5800 K. Historically, the first measurements of this solar irradiance spectrum were taken at the Earth's surface or at altitude to mitigate the effects of the atmosphere. In particular, Labs and Neckel took measurements of the solar irradiance spectrum at the Swiss research station of Jungfrau (3580 m) in 1960. More recently, the spectrum was obtained by combining satellite measurements, from high altitude airborne measurements and numerical modeling. Figure 1.8(b) shows the spectral variations of solar irradiance at the top of the atmosphere and on the Earth's surface. Note that not all of the solar flux reaches the surface, some may be absorbed by atmospheric gases or dispersedby atmospheric molecules. Planck's curve corresponding to a black body radiating at 5800 K is also reported and does not correspond exactly to that emitted by the Sun, highlighting that it is not exactly a black body. Approximately half of the radiation is distributed in the visible light region ($0.38\ \mu m \geq \lambda \geq 0.78\ \mu m$) and half in the infrared ($\lambda \geq 0.78\ \mu m$), with approximately 1% in the ultraviolet.

Integrating the solar function across the spectrum defines the total solar irradiance. For Earth, the solar function, often called the solar constant and denoted F_0, expresses the solar flux received over an area of 1 m^2 outside of the atmosphere, at a distance of 1 AU (Astronomical Unit, representing the mean distance between the Earth and the Sun) from the Sun and perpendicular to the incident rays from the Sun. The actual known value for F_0 is equal to 1360.8 ± 0.5 $W.m^2$. The term "solar constant" is commonly used, even though the solar flux density at the top of the atmosphere varies with time. In fact, the Sun is a dynamic star and has sunspots on its surface whose surface temperature is lower than that of their environment and varies over time (sunspots were first observed at the beginning of the 17th Century by Galileo). Solar activity therefore alternates between maximum and minimum peaks approximately every 11.2 years; this is known as the Schwabe cycle, named after the German astronomer who discovered it (1843). The fluctuation in the total solar flux emitted is approximately ± 1 W.m^2. Moreover, as the Earth's trajectory is elliptical, the distance between the Earth and the Sun varies during a revolution by approximately ± 3.3 %. This distance can be calculated using the simple relation given by Duffie and Beckman (1991), which determines the corrected value of the solar flux received F such that $F = F_0\left[1 + 0.033\cos\left(\frac{360n}{365}\right)\right]$, with n the number of days (between 1 and 365, n = 1 corresponding to the 1st January) and the angle is expressed in degrees. Note that a more precise formula (at ± 0.01 %) was established by Spencer in 1971. Finally, the flux of energy received by the Earth also depends on the periodic parameters of Earth's orbit. The consequences of these orbital parameters on Earth's climate were discovered and described by Milutin Milankovitch (1941) in his theory called the astronomical theory of climate. This explains the variations in sunlight on the Earth's surface as the Earth moves around the Sun and in particular the occurrence of glaciations of the Pleistocene. Three fundamental independent parameters characterize the position of the Earth in relation to the Sun: the eccentricity (characterizing the shape of the Earth's orbit, with a cycle of approximately 100000 years), the axial tilt of the Earth (or obliquity of the ecliptic, which gives the angle between the Earth's rotational axis and the axis perpendicular to the orbital plane, with a cycle of approximately 40000 years) and the axial precession (also called precession of the equinoxes, gradual change in orientation of the Earth's rotational axis, with a cycle of approximately 22000 years).

The Earth's climate system, heated by the Sun, emits in the thermal infrared range. Figure 1.9 presents numerical simulations of spectra of $L_\nu(T)$ emitted in the thermal infrared, like those seen by a spectrometer from space when observing the Earth's surface at the nadir (Figure 1.9(a)) or from the ground looking at the sky at the zenith (Figure 1.9(b)). These simulations were carried out for a typical atmosphere at mid-latitudes, including atmospheric gases with a surface similar to a black body with a temperature equal to 288 K. For comparison, Planck's curves have been added for temperatures between 150 and 300 K. Note that in Figure 1.9(a), the simulated radiances at the top of the atmosphere (TOA) in the region between 800 and 1200 cm^{-1} correspond to brightness temperatures of approximately 290 K. In fact, this spectral interval corresponds to a region of the spectrum relatively transparent to radiation, for which gaseous absorption is low overall. The radiance observed is therefore the sum of that emitted by the surface, which is poorly absorbed by the atmosphere and that emitted by the atmosphere at a low intensity due to the transparency of the latter. Equivalent brightness temperatures simulated at TOA (represented by dotted lines) are therefore close to the surface temperature. However, note the absorption peak observed in Figure 1.9(a) and (b), for wavenumbers between 1000 cm^{-1} and 1100 cm^{-1}, which correspond to the absorption band of ozone (approximately 9.6 µm). On the other hand, the radiances simulated for this region of the spectrum for an observation from the ground (BOA) correspond to low brightness temperatures (Figure 1.9(b)). In fact, the main target this time is space, at a radiative temperature close to 2.7 K, and the brightness temperature of the atmosphere simulated from the ground is less than 200 K. This region of the spectrum, called the infrared atmospheric window, is therefore paramount in remote sensing because it allows a space sensor to effectively probe the surface or lower layers of the atmosphere or allows a sensor placed on the ground, such as an infrared telescope, to observe the space. As the absorption of atmospheric gases is greater on either side of the atmospheric window, the simulated radiances in these spectral regions are higher at the BOA (Figure 1.9(b)) and lower at the TOA (Figure 1.9(a)), the layers of the atmosphere observed at these wavenumbers being lower or higher in altitude, respectively. For example, the effect of the strong absorption band of carbon dioxide is visible between 600 and 700 cm^{-1} (approximately 15 µm).

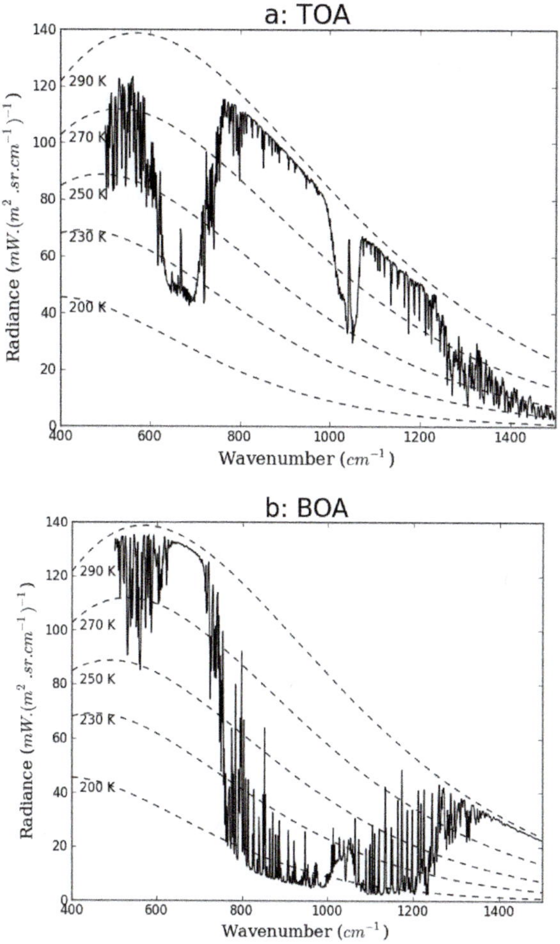

Figure 1.9. *Atmospheric radiance spectra (solid lines) as seen from top of atmosphere; a) TOA, or from the bottom of the atmosphere; b) BOA, simulated for a standard mid-latitude atmosphere and over a black surface with a temperature of 288 K. Planck function curves corresponding to different blackbody temperatures are superimposed (dashed lines, from 200 to 290 K)*

1.4.4. *Reflection and emission of radiation by a surface*

The contribution of radiation emitted or reflected by the surface of the Earth to the signal measured at the top of the atmosphere is large in the infrared region. In the solar spectrum, the incident radiation may be partially reflected, the rest being absorbed. The reflection of radiation by a surface occurs either in a specular manner off a relatively flat surface (calm sea, lake) or diffusely off a very irregular surface (vegetation cover, heavy seas, etc.) (Figure 1.10 left). In the first case, we describe the reflection using a factor given by Fresnel's law; in the second, we introduce the spectral reflectivity factor α_λ, or spectral albedo, which is the ratio of the reflected flux F_λ^r and the incident flux F_λ^i. We therefore define the reflectivity factors α_λ and spectral absorptivity a_λ, such that $\alpha_\lambda + \alpha_\lambda = 1$. Some surfaces reflect radiation isotropically, regardless of the wavelength λ, and are known as Lambertian. However, many surfaces are characterized by their non-Lambertian character. Under a directional incident irradiance $\mathbf{s}_i$, we quantify the reflection using the Bidirectional Reflectance Distribution Function (BRDF) $\rho_\lambda(\mathbf{s},\mathbf{s}_i)$,which is the ratio of the radiance reflected in the solid angle about the reflection direction $\mathbf{s}$, $dL_\lambda^+(s)$,on the irradiance of the flat surface whose normal has an angle θ_I with the direction of incidence:

$$\rho_\lambda(s,s_i) = \frac{dL_\lambda^+(s)}{L_\lambda^-(s_i)\cos\theta_i d\Omega_i}. \qquad [1.11]$$

Note that if the surface is Lambertian, the BRDF is constant and it is called reflectance ρ_λ, which is distinguished from albedo due to its directionality. In this case, we obtain the relation between the radiance reflected I_λ^+ and the incident flux F_λ^i:

$$L_\lambda^+ = \frac{\rho_\lambda F_\lambda^i}{\pi} = \frac{\rho_\lambda \cos\theta_i E_\lambda^i}{\pi} = \frac{F_\lambda^r}{\pi}. \qquad [1.12]$$

Figure 1.10 (right) shows some examples of Lambertian reflectance spectra (expressed in %) for different types of surfaces.

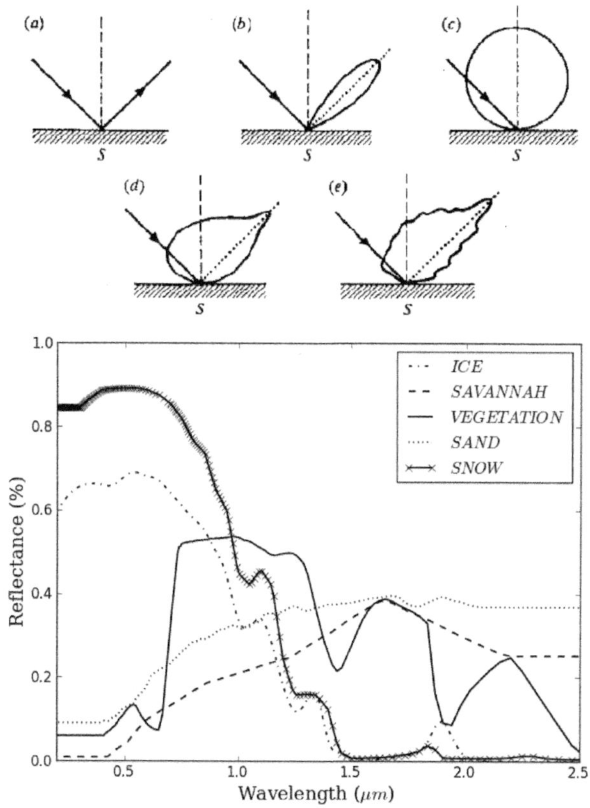

Figure 1.10. *(Bottom) Schematic representation of different types of surface reflection: a) Specular, b) quasi-specular, c) Lambertian, d) quasi-Lambertian, e) complex. (Top) Lambertian reflectance spectra $\rho_\lambda(\%)$ as a function of wavelength for different types of surfaces*

In the thermal infrared, a terrestrial surface is characterized at a given wavelength by its spectral emissivity ε_λ, defined as the ratio of its spectral radiance L_λ to that of a black body B_λ at the same temperature: $\varepsilon_\lambda = \dfrac{L_\lambda(T)}{B_\lambda(T)}$. For a black body $\varepsilon_\lambda=1$; for any body, $\varepsilon_\lambda < 1$.

A body for which the emissivity is independent of wavelength is classed as a gray body. Some concrete examples of average emissivity and albedo are shown in Table 1.5.

Medium	Emissivity	Albedo	Medium	Emissivity	Albedo
Mineral surface	0.85–0.95	0.1–0.55	Sea surface (sun near zenith)	0.94–0.98	0.02–0.05
Asphalt	0.95–0.97	0.05–0.2	Liquid water	0.99–1	0.05–0.2
Urban zone	0.85–0.87	0.1–0.2	Ice	0.96–0.98	0.25–0.7
Green grass	0.9–0.99	0.16–0.26	Fresh snow	0.99–1	0.75–0.95
Forest	0.95–0.97	0.05–0.2	Old snow	0.82	0.4–0.7
Sun – Earth	0.9–0.98	0.05–0.2	Ice clouds	0.1–0.9	0.1–0.7
Sand – desert	0.84–0.96	0.2–0.4	Liquid clouds	0.25–1	0.3–0.9

Table 1.5. *Characteristic values of average emissivity in the thermal infrared (between 8 and 12 μm) and of average albedo in the visible spectrum for different surface types*

1.5. Radiation budget of the climate system

1.5.1. *Radiative balance of the atmosphere*

Solar radiation is the main energy source in the Earth's current climate system; other energy sources, such as from geothermal energy, are negligible. It therefore plays a key role in atmospheric processes and is naturally behind most physical phenomena (turbulence, convection, evaporation, etc.). Variations in the radiation energy received at the surface of the Earth thus have a direct influence on the radiation budget of the atmosphere and thus on climate. The Earth's climate system receives on average about 1361 $W.m^{-2}$ at the top of the atmosphere. Assuming that the Earth is spherical, the average energy received at the surface is close to 340 $W.m^{-2}$. Part of this radiation is reflected by atmospheric components (molecules, aerosols and clouds) and the Earth's surface; this fraction of reflected radiation is the planetary albedo and is around 30%. The other part is absorbed by the system, primarily by the surface, and helps heat it. As the system is heated, it emits energy into space as infrared radiation at wavelength of between 4 and 100 µm. It is thought that radiative equilibrium occurs in this system when the amount of energy received is equal to the amount of energy emitted or reflected. Therefore, the radiation budget is the difference between the energy actually received by the Sun, and that actually lost by reflection or thermal emission to space. This radiative equilibrium is reached at a global level and is shown in Figure 1.11. However, there is a large inconsistency at the local level, especially with latitude: there is a surplus of energy at low latitudes and an energy deficit beyond 45° latitude (Figure 1.12). This imbalance results in the transfer of energy on a global scale, from low to high latitudes, and is responsible for atmospheric and oceanic dynamics. On a global scale, the radiation budget is only balanced at mid-latitudes.

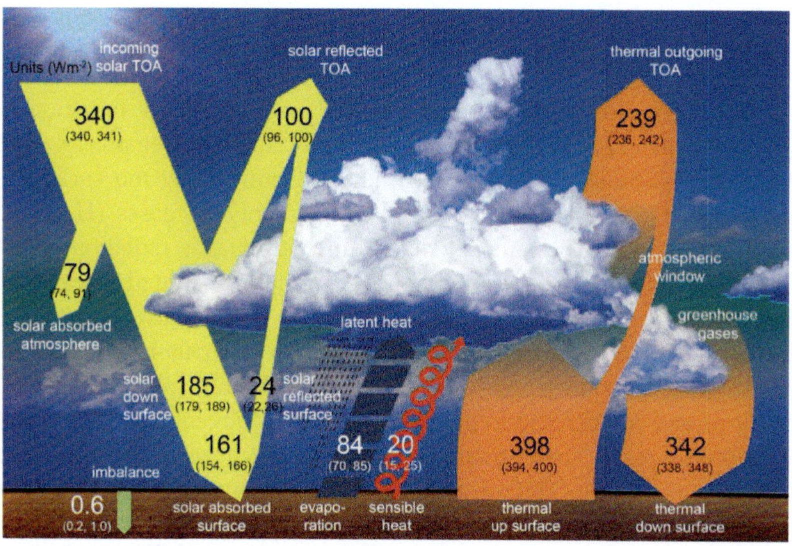

Figure 1.11. *Global mean energy budget under present day climate conditions. Numbers state magnitudes of the individual energy fluxes [W.m^{-2}], adjusted within their uncertainty ranges to close the energy budgets. Numbers in parentheses attached to the energy fluxes cover the range of values in line with observational constraints. Source: [HAR 13]*

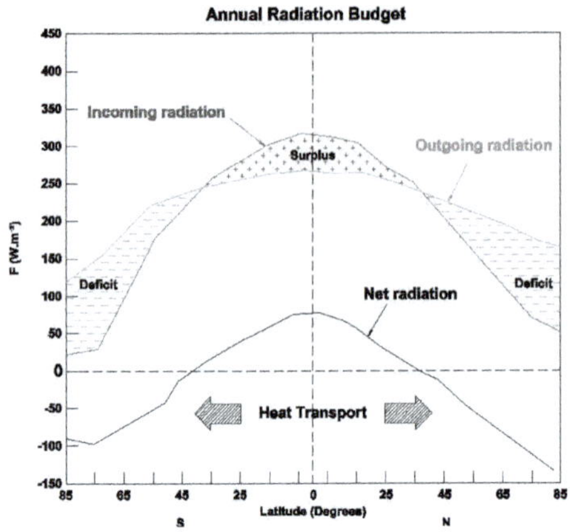

Figure 1.12. *Annual radiation budget [W.m^{-2}] of the atmosphere (net radiation) as a function of latitude, defined as the difference between radiation gained (incoming radiation) and that lost (outgoing radiation) by the atmosphere*

1.5.2. *The greenhouse effect and parasol effect*

The Earth's atmosphere absorbs little incoming solar radiation in short wavelengths (Figure 1.8(b)). However, with regard to long wavelengths, most of the infrared radiation emitted by the surface is absorbed by the atmosphere (Figure 1.9), mainly by gases (H_2O and CO_2) and clouds. The atmosphere, in turn, emits this radiation to the surface or into space (Figure 1.13). The radiation emitted toward the Earth's surface warms the Earth: this is the greenhouse effect. The current mean surface temperature of the Earth is about 15°C; it would be -18°C without the greenhouse effect. For an atmosphere consisting solely of gas, an increase in the concentration of greenhouse gases, and therefore a greater absorption of radiation in the infrared, increases the surface temperature. The intensity of the greenhouse effect depends on the atmospheric concentration of gases, their absorption capacity and their residence time in the atmosphere. Also, note that there is only a greenhouse effect if the temperature of emission into space is lower than that at the surface, which is the case for Earth's atmosphere. Consequently high altitude clouds, such as cirrus, produce a significant greenhouse effect. In addition, there is no saturation effect for the greenhouse effect in the case of carbon dioxide. Indeed, the absorption of infrared radiation by CO_2 in the entire atmospheric column is almost at its maximum in current conditions and it only slightly depends on concentration variations of this gas. It is thought that this absorption is saturated. However, it is important to take into account the variation in the radiation emission altitude with CO_2 concentration. The existence of a vertical gradient of atmospheric temperature, associated with the absorbency saturation of infrared radiation by CO_2, explains the increase in the greenhouse effect when its concentration increases, as well as associated warming climate. However, it is worth noting that, by analogy and contrary to popular belief, blocking infrared radiation using glass is not the primary mechanism that explains how a greenhouse functions. Therefore, radiative forcing is the scientific terminology adopted by the IPCC to describe the influence of atmospheric components that block infrared radiation in Earth's radiation budget.

In parallel, with the exception of absorbent particles such as soot, aerosols and hydrometeors predominantly scatter radiation and help cool the climate system: this is the parasol effect (Figure 1.13). For example, volcanic dust emitted into the stratosphere during large volcanic eruptions can alter the surface temperature by approximately -0.5 to -1.5 K. These variations can be measured over a period of several months to several years, depending on the residence time of aerosols in the stratosphere. The parasol effect can therefore, in theory, offset the increase in temperature caused by the greenhouse effect. An exact estimate of the greenhouse and parasol effects to assess the magnitude, speed and geographical distribution of global warming is a current challenge.

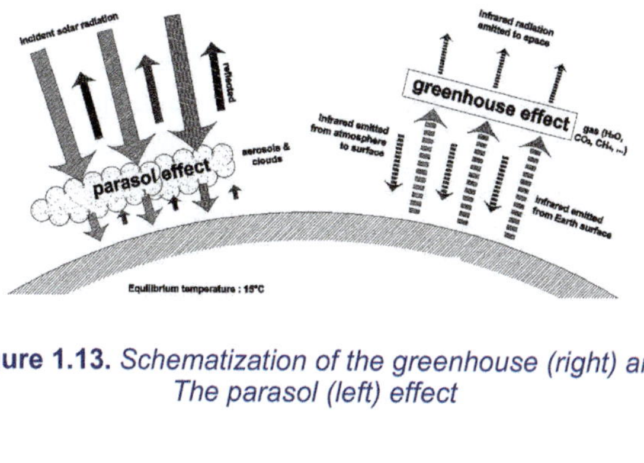

Figure 1.13. *Schematization of the greenhouse (right) and The parasol (left) effect*

1.5.3. *Radiative forcing of atmospheric components*

Radiative forcing [$W.m^{-2}$] is a disruption to the radiation budget of the climate system by external factors, such as anthropogenic emissions of greenhouse gases or aerosols. The term "forcing" therefore indicates a change in the balance between radiation received by the system and the loss by transmission or reflection. If radiative forcing is positive, there is an increase in energy in the system and therefore a warming thereof; in the opposite case, if it is negative, there is a cooling of the system. Quantitatively, radiative forcing ΔF is written as the difference in net flux at the top of the atmosphere, calculated over the whole spectrum, in response to an external disturbance: $\Delta F = F^{net}_{perturbation} - F^{net}_{no-perturbation}$. The estimated radiative forcing allows us to quantify the impact many factors have on global warming. However, to fully assess the contribution of one component to the greenhouse or parasol effect over time, we must consider its impact on spectral radiation, its concentration and its residence time in the air, and whether this component can accumulate in the atmosphere. This contribution is difficult to quantify, especially since the effects are generally nonlinear. For example, an estimate of the impact of greenhouse gases on the radiation budget is given by the global warming potential (GWP), calculated relative to a reference gas (here CO_2) and defined as:

$$GWP_x = \frac{\int \Delta F_x(t)dt}{\int \Delta F_{CO_2}(t)dt} = \frac{RE_x \tau_x \left(1-\exp[-t/\tau_x]\right)}{RE_{CO_2} \tau_{CO_2} \left(1-\exp\left[-t/\tau_{CO_2}\right]\right)} \frac{M_{CO_2}}{M_x}, \quad [1.13]$$

where ΔF_x is the radiative forcing, τ_x is the lifetime in the atmosphere and M_x is the molar mass; the index x refers to the gas considered. The radiative efficiency RE [$W.m^{-2}.ppb^{-1}$] is defined as the derivative of radiative forcing and quantifies the rate of change in radiative forcing in terms of mass. Some examples of GWP are given in Table 1.1. Note that the global warming potential of water vapor cannot be accurately calculated due to its direct link between concentration of water vapor and air temperature.

Figure 1.14 shows an estimate, made by the IPCC, of radiative forcing since the beginning of the industrial revolution, as well as their uncertainty. The forcing caused by greenhouse gases is about +3.0 W.m^{-2}, a value estimated with an accuracy of 10%. Forcing by atmospheric aerosols is overall negative but with a much greater level of uncertainty. In total, the estimates show anthropogenic forcing to be positive overall at approximately +2.3 W.m^{-2}, which equates to global warming. Figure 1.14, however, shows a high level of uncertainty in the total radiative forcing due to clouds and aerosols. For a better estimate of the overall radiative forcing, the physical and microphysical properties of aerosols and clouds must be better understood.

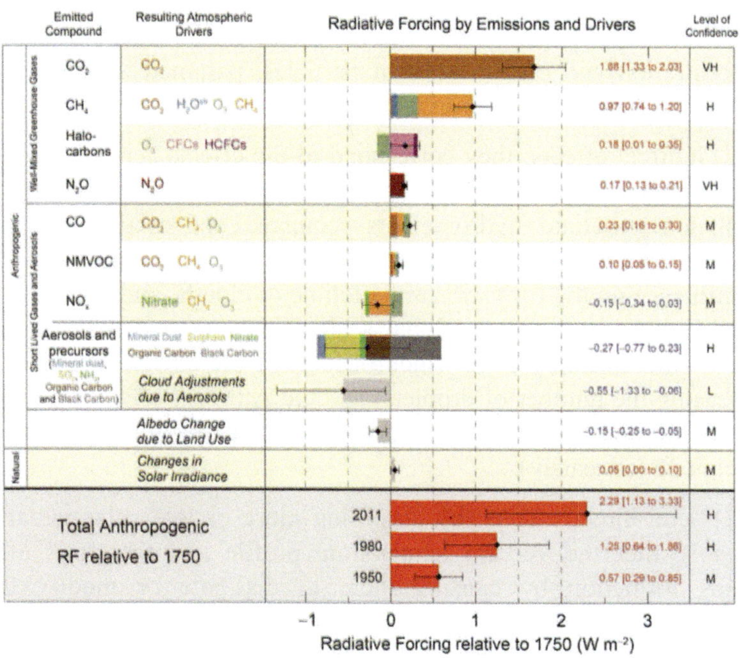

Figure 1.14. *Radiative forcing by concentration change between 1750 and 2011 with associated uncertainty range. Source: [IPC 13] (see color section)*

1.5.4. *Impact of aerosols on climate*

Already in 2001, the IPCC had highlighted the importance of better understanding aerosols and their influence on radiation. In 2007, the IPCC report dedicated a whole chapter to aerosols and their interactions with the climate. Their effects are difficult to quantify because the distribution of aerosols is very variable in space and time and the interaction processes with radiation and other components in the climate system are complex. The interest in aerosols is increasing particularly those of anthropogenic origin. The influence of aerosols can be classed into three effects:

1) Direct effect: through the direct interaction with solar and terrestrial radiation, atmospheric aerosols modify the Earth's radiative equilibrium via the greenhouse and parasol effect. Globally quantifying the direct effect of aerosols is still tricky since their radiative forcing is highly variable (from about 10 $W.m^{-2}$ to approximately 50–80 $W.m^{-2}$ in extreme cases) and depends on the concentration and composition of particles, parameters with a high spatial and temporal variability.

2) Indirect effects: they correspond to the effects aerosols have on cloud properties, which in turn affect the climate. Indeed, the formation of cloud hydrometeors requires condensation nuclei to activate the condensation process. The aerosol concentration can have an influence on the formation and lifetime of clouds, and thus alter the albedo at the regional or global scale. Note the two main indirect effects: the Twomey effect (modification of the albedo of clouds by increasing the number of droplets) and the Albrecht effect (prolonged cloud lifetime due to the presence of smaller droplets, which precipitate less easily).

3) Semi-direct effects: by absorbing more or less solar radiation, aerosols alter the vertical temperature profile and therefore, affect clouds formation by causing them to dissipate or modify their geographical extension. This effect is linked to the type of surface that aerosols cover. For example, a layer composed of absorbing aerosols can heat up under the effect of solar radiation and halt or limit the development of a cloud lying beneath. Moreover, thermal gradients

associated with the vertical distribution of absorbing aerosols may modify the vertical stability of the atmosphere.

In general, indirect and semi-direct effects are still poorly estimated due to their relatively complex nature.

1.5.5. *Impact of clouds on climate*

Clouds have a significant influence on radiation and their role on the climate is therefore paramount: without them, the planetary albedo would be approximately halved. This influence manifests, according to their nature and altitude, as the greenhouse effect, or on the contrary, the albedo effect. The amount of incoming solar energy is lower under a thick cloud where the radiation is directly reflected back into space. Conversely, at night, cloud cover limits the temperature drop, by absorbing infrared radiation emitted from the surface and returning it. Regardless of the cloud, the main effect depends on its temperature (and therefore its altitude) and its optical and microphysical properties that determine its ability to reflect solar radiation and absorb infrared radiation. In particular, the cloud thermodynamic phase (liquid water or ice), as well as the size of cloud particles are decisive. Moreover, the spatially largest clouds and those most persistent over time have the greatest effect on radiation.

It is generally accepted that low altitude clouds, usually warmer and denser, effectively reflect sunlight and thus tend to cool the atmosphere. As an example, stratocumulus clouds are not very thick, but very extensive and more persistent because they do not contain much water and so do not precipitate. Since they are low, their temperature is slightly different from the surface temperature and their greenhouse effect is limited. However, their reflectivity exceeds 40% and they therefore have a strong albedo effect (parasol effect). Conversely, high altitude clouds are coldest and tend to warm the atmosphere. This is the case of cirrus clouds, which are very broad and cover about 20% of the Earth's surface. They are composed of ice crystals, which strongly absorb infrared radiation, and with a temperature lower than –40°C, this intensifies their greenhouse effect.

Moreover, cirrus clouds are relatively transparent to solar radiation and their albedo effect is low. The radiative balance estimated using satellites helped assess these effects, with an annual average of approximately – 48 Wm^{-2} for the albedo effect and 28 Wm^{-2} for the greenhouse effect. Overall, clouds on average have a slight cooling effect on the Earth's climate system. If they became transparent, the radiation balance would increase by about +20 Wm^{-2}. However, there is still a lot of uncertainties when estimating these effects on a global scale. The climate system is sensitive to changes in the vertical structure, spatial distribution and occurrence of clouds. The estimated radiative forcing due to a doubling of CO_2 could be offset by a relative increase of 15–20% of the amount of low altitude clouds.

1.5.6. *Climate sensitivity*

An additional difficulty when assessing the role aerosols and clouds have on climate is to estimate climate sensitivity, that is to say, the change in atmospheric temperature in response to radiative forcing. In fact, if climate change alters one of the components of the climate system and this change in turn modifies the climate, it is called feedback. It is called negative feedback if it opposes the change and positive if it amplifies it. A classic example of positive feedback is water vapor, the main greenhouse gas, which reacts very quickly to climate change: the hotter it gets, the more water vapor there is in the atmosphere, which increases the greenhouse effect and so reinforces climate warming. However, an increase in atmospheric water vapor can also cause overall cloud cover to increase, which as we have seen, tends to cool the climate via the albedo effect. Clouds could therefore be a negative feedback on global warming.

The physical and chemical principles allow us to calculate the direct influence of factors that affect the Earth's climate. However, in order to estimate the outcome of many couplings occurring in the climate system, numerical modeling itself is the only scientific approach. It is, however, not perfect because one cannot completely model a system as complex as the Earth and feedback remains the largest source of uncertainty when assessing climate change [IPC 13].

It is therefore essential to have monitoring systems at the global and regional level, with the required accuracy and reliability to measure the long-term variability of clouds and radiation budget and to have long time series as a reference for models and simulations to validate them. Observing the climate system is essential in order to better understand how clouds respond to global warming, as well as anthropogenic aerosol emissions.

1.5.7. *Observation of radiative budget*

Measuring the radiation budget and its effects on the mean surface temperature of the Earth is complex because it must be done globally. There are many thousands of automatic meteorological stations worldwide used to measure the surface temperature of the Earth. Their distribution is, however, not uniform, with many in North America and Europe and fewer in Africa and South America, and almost none over the oceans. These measurements are even rarer at altitude (atmospheric radiosondes). Space observations in turn allow the radiation balance at the global level to be estimated. This estimate, however, is tricky as the satellites do not directly measure the temperature and radiative fluxes, but the radiance emitted or reflected by the Earth-atmosphere system in a given direction and for certain wavelength ranges. The use of satellites, however, allowed the first global estimates of the radiation budget of the Earth with simple instruments onboard Explorer 7 (1959) then Tiros (1960). The series of Nimbus satellites then provided, between 1985 and 1990, the foundations ofthe Earth Radiation Budget Experiment (ERBE). Following the ERBE, other radiometers embarked on Scanner for Radiation Budget (ScaRaB) missions between 1994 and 1999. Currently, sensors of the space program Clouds and the Earth's Radiant Energy System (CERES), were deployed on the TRMM satellite, Terra and Aqua of NASA's Earth Observing System (NASA: National Aeronautics and Space Administration), which provide measurements of the radiation budget and thus ensure the continuity of the radiative flux time series measurements from space. Figure 1.15 shows an overall estimate of the radiation budget at the top of the atmosphere, obtained after processing CERES satellite data for the month of September 2012 and reveals the high spatial variability of the radiation budget. As one might expect, we can notice that this budget is negative toward the poles, largely positive near the equator, since it receives more solar energy, and approximately zero at mid-latitudes.

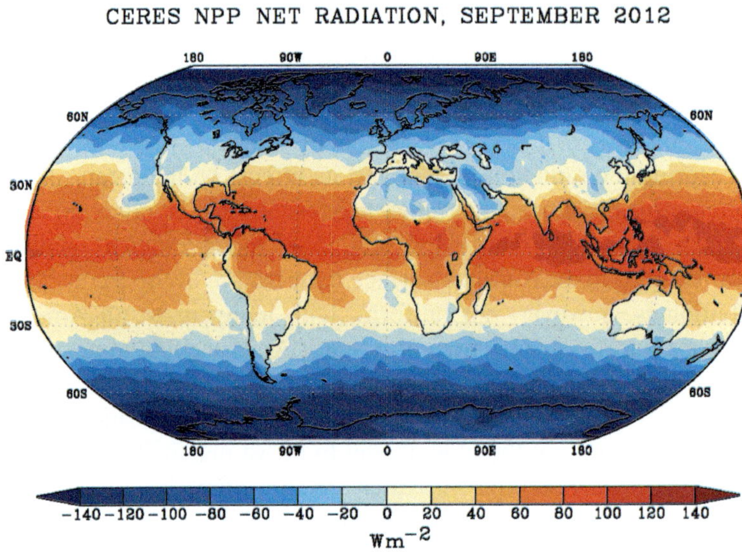

Figure 1.15. *Global Image of Net Radiation at Top-of-Atmosphere from CERES Instrument on NASA Suomi NPP Satellite for September 2012 (Instrument: CERES/NPP; Release Date: 12/7/2012). Source: NASA/CERES website http://ceres.larc.nasa.gov/press_releases.php (see color section)*

1.6. For further information

[ACK 13] ACKERMAN S.A., KNOX J.A., *Meteorology: Understanding the Atmosphere*, Jones & Bartlett Learning, fourth edition, 2013.

[BOU 15] BOUCHER O., *Atmospheric Aerosols – Properties and Climate Impacts*, Springer, Netherlands, 2015.

[HOU 09] HOUGHTON J., *Global Warming: The Complete Briefing*, Cambridge University Press, fourth edition, 2009.

[PET 06] PETTY G.W., *A First Course in Atmospheric Radiation*, Sundog Publishing, Madison, Wisconsin, 2006.

[WAL 06] WALLACE J., HOBBS P., *Atmospheric Science*, second edition: An Introductory Survey, Academic Press, second edition, 2006.

[WAN 13] WANG P.K., *Physics and Dynamics of Clouds and Precipitation*, Cambridge University Press, 2013.

2

Instrumentation and Sensors

Remote sensing is a key element in the study of the Earth-ocean-atmosphere system, since it represents the only observation method at the regional to global scale. The first civil Earth observation applications from space began in the 1960s and were dedicated to weather monitoring. For the first time, the development and movement of atmospheric systems could be monitored continuously by providing atmospheric humidity and cloud cover observations across the entire globe. These efforts in weather observation continued with geostationary instruments such as GOES (United States), GMS (Japan) and METEOSAT (Europe) from the 1970s, followed by NOAA polar orbiting satellites deploying the AVHRR sounder. It was at this time that space missions dedicated to the observation of continental and marine surfaces emerged, in particular with the LANDSAT, SPOT and Nimbus 7 satellites. Consequently, many applications in the fields of cartography, habitat management and agriculture or geology have benefited from satellite observations.

However, it was only in the 1980s that the first missions, dedicated to studying the radiation budget of the Earth-atmosphere system began with instruments such as ERBE, CERES, and ScaRaB. It was also during this period that studies of atmospheric composition began (pollution, trace gases) with the missions ATMOS and MAPS. Then in the 1990s, the IMG sounder, despite its short lifespan, confirmed that hyperspectral sounders could quantify the gaseous composition of the atmosphere. Since then, monitoring changes in atmospheric composition has benefitted from many broadband infrared instruments such as MODIS and SEVIRI or high spectral resolution instruments such as TES, ACE-FTS, IASI, and TANSO-FTS.

In this chapter, we will discuss the technical aspects of obtaining measurements using satellite remote sensing. We will also present the specifics of satellite platforms, including the types of orbits and viewing modes together with their uses in atmospheric studies. These characteristics of satellite remote sensing are common to all the on-board instruments regardless of their wavelength range or application. Finally, we will concisely present the main methods used in optical infrared remote sensing.

2.1. Platforms, satellites and sensors

2.1.1. *Types of orbits*

An orbit is the closed trajectory that a satellite takes around a center of attraction, in this case, the Earth. This trajectory is governed by the three Kepler laws that apply to the movement of an object orbiting a celestial body. The orbit generally has the shape of an ellipse with one of its focus points at the center of the Earth, around which the satellite revolves. The latter is characterized by: perigee and apogee, which are the distances from the closest point (r_p) and the farthest point (r_a) to the Earth, respectively (Figure 2.1), as well as the eccentricity $e = (r_p\text{-}r_a)/(r_p\text{+}r_a)$, which defines the extension of the ellipse and whose value is between 0 and 1, with a value of zero corresponding to a circular orbit. The orbits of artificial satellites are generally classified according to their distance from Earth:

– low orbit lies just above the Earth's atmosphere at an altitude of less than 2000 km. It is mainly used for satellite instruments dedicated to Earth observations;

– average orbit generally culminates at an altitude of 20000 km. At this distance away from the Earth's atmosphere, the satellite orbit is very stable. In addition, signals from the satellite can be received over a large part of the Earth's surface. It is therefore the type of orbit used with navigation instruments such as GPS;

– high orbit is characterized by an altitude greater than 30000 km. The geostationary orbit is the most common high orbit used predominantly for earth observations. However, there are other high orbits as the Molniya orbit, which is a highly eccentric orbit with an apogee of 40000 km and a perigee of 500 km. The latter can cover the high latitudes and is particularly used by Russia for communication satellites.

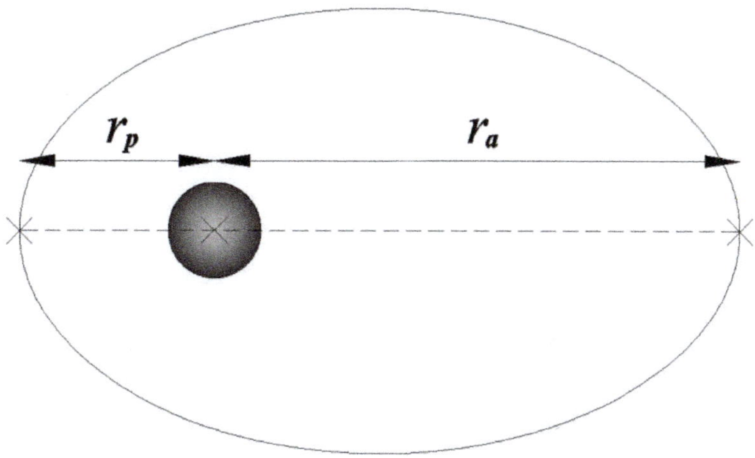

Figure 2.1. *Example of an elliptical orbit about a center of attraction. The distances r_p and r_a denote the perigee and apogee, respectively*

The orbit of a satellite is determined by the ability of the sensors it carries and its mission objectives. Satellites used to observe the Earth are divided into two main categories: low orbit and geostationary.

2.1.1.1. *Geostationary orbits (GEO)*

Satellites in geostationary orbit (GeO) rotate around the Earth on a circular trajectory at an altitude of about 36000 km from the surface in the equatorial plane (deduced from Kepler's 3rd law). At this altitude, the satellite's orbital period is exactly the same as the rotation period of the Earth (geosynchronous), so they appear stationary relative to a point on the globe. GeO satellites can therefore continuously observe the same portion of the Earth's surface (Figure 2.2), which makes them particularly suitable for measurements requiring a good sampling period. However, as this orbit is situated on the equatorial plane, it does not cover areas that are located at high latitudes (>80°). At present, only satellite communications and weather observation systems (e.g. MSG) use such orbits. However, one platform (Meteosat Third Generation) comprising at least one infrared instrument, IRS, dedicated to chemistry and climate variables will soon be launched by ESA/Eumetsat. This instrument should enable the observation of the whole of Europe and North Africa. Similar projects are being prepared by NASA and Jaxa, to allow equivalent observations of the American and Asian continents.

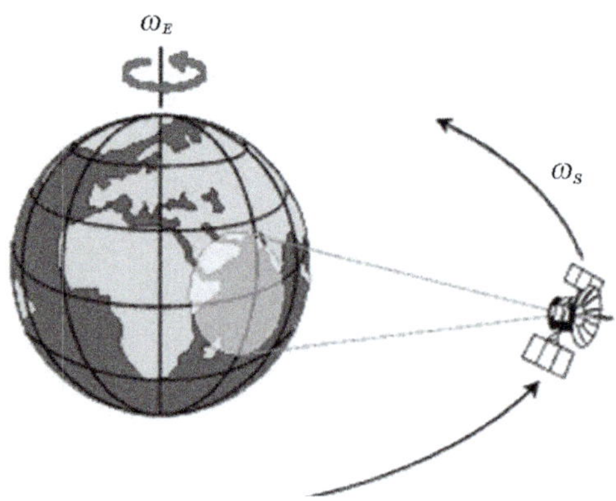

Figure 2.2. *Observation of the Earth by a satellite in a geostationary orbit. In this case, the angular velocities of the Earth ω_E and the satellite ω_S are equal*

2.1.1.2. *Low Earth orbits (LEO)*

Low orbit platforms (LEO) are generally located on a polar orbit, which is a circular path at an altitude of less than 1000 km. Its inclination (angle between the plane containing the satellite and that of the equator) close to 90° is passed near the poles (Figure 2.3). This trajectory (North-South), combined with the rotation of the Earth

(West-East), has the advantage of being able to regularly observe the entire Earth. Most LEO satellites are helio-synchronous and their orbital plane is slightly greater than 90° (near-polar), and in this case retain an angle consistent with the Earth-Sun direction; in this way, a given region of the world is observed at the same local solar time. This orbital characteristic ensures similar solar illumination conditions are observed for the same region. The satellite platforms that are in near-polar orbits, thus moving north on one side of the Earth, and to the south in the other half of their orbit, known as an ascending and descending orbit, respectively. If the orbit is helio-synchronous, the ascending orbit of the satellite is usually the shaded side of the Earth, while the descending orbit is in the sunlit side. Thus, only active sensors, which have their own light source or passive sensors, which record the energy emitted by the planet, that is to say in the thermal infrared range of the spectrum, can continuously acquire data.

Figure 2.3. *Observation of the Earth by a low-orbit satellite*

2.1.2. *Characteristic parameters of satellites*

Regardless of the type of orbit or application area, artificial satellites are characterized by a certain number of common parameters, which are defined below.

– *ground track*: all the points on the surface vertically below the satellite. The ground track combines the motion of the satellite and the rotation of the Earth. Figure 2.4 shows the ground track of the satellite Metop represented on a world map;

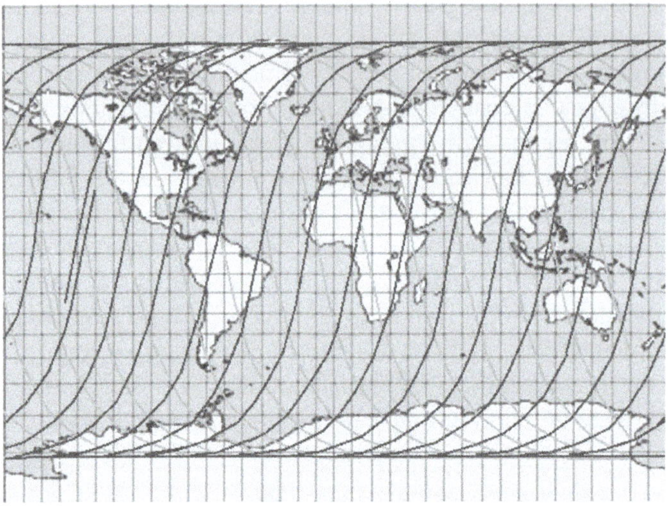

Figure 2.4. *Ground track of instrument IASI/Metop*

– *orbital cycle*: this is the total number of revolutions made by a satellite around the globe in order to return above its starting point;

– *revisit period*: this corresponds to the period required for the instrument on board the satellite platform to observe the same point of the globe again. The revisit period can be likened to the temporal resolution of the measurement;

– *IFOV* (*Instantaneous Field of View*): this is the solid angle [mrad] with which the instrument sensor is sensitive to electromagnetic radiation. The IFOV and the distance from the source define the spatial resolution of the instrument.

2.1.3. *Geometry of lines of sight*

2.1.3.1. *Nadir*

The nadir line-of-sight corresponds to the vertical downward direction (Figure 2.6), i.e. the observation points on the surface of the

Earth that are directly below the satellite trajectory. When observing the reflected light, the sensitivity will depend on the position of the Sun, absorption characteristics, reflection and scattering of atmospheric particles and surface properties (albedo, BRDF). When observing thermal emissions, the sensitivity depends on the thermal contrast the surface and the atmosphere, as well as surface properties (emissivity). The thermal inertia of the atmosphere is greater than that of the seas and oceans, itself larger than that of continental surfaces and the contrast is larger during the day than at night. Thus, the sensor sensitivity in the thermal IR will be greater during the day above the continents and lower at night over the oceans.

In the nadir viewing mode, some sensors undertake scanning observations. Such measurements may be taken using three different techniques:

– using a mobile (or turning) mirror at the input of the instrument that guides the field of view as the satellite moves (e.g. AVHRR sensor of *NOAA* satellites);

– an association of thousands of detectors side by side, known as "push-broom" (e.g. *Spot* satellites);

– rotation of the satellite itself and moving of the telescope's line of sight (e.g. *Meteosat* satellite sensor).

The extent of spatial coverage via scanning perpendicularly to the ground track is called "swath". It is represented in Figure 2.5 for the instrument IASI/Metop and can be compared to the ground track in Figure 2.4.

Thus, the instruments that perform scanning measurements can observe the same surface area before and after the passages of the main orbit, corresponding to the nadir, which allows a much shorter revisit time than the orbital cycle. The revisit time is different for each instrument and proves to be an element that determines the spatiotemporal capacity of these observations. In general, near-polar orbiting satellites more frequently cover high latitude regions than equatorial zones. This increased coverage is due to enlargement, toward the poles, of the overlap between adjacent swaths.

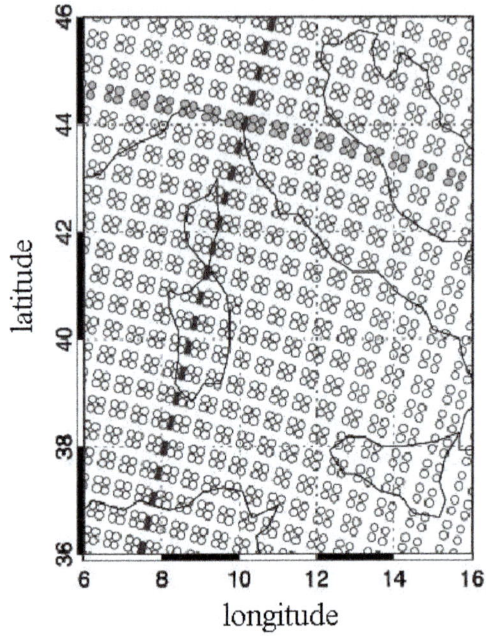

Figure 2.5. *Spatial coverage of the instrument IASI/Metop. The black dotted line represents the ground track. All the gray circles (ground surfaces corresponding to IFOV) represent the swath*

At the nadir line-of-sight, the horizontal spatial resolution (also commonly called pixel size) of observations is represented by the area on the ground that is seen by the instrument at a given time. This is determined by the IFOV and the distance between the surface and the satellite. Note that if the observation is not made exactly on the nadir (i.e. nonzero observation angle), the IFOV becomes oblique, which deteriorates the spatial resolution and changes the optical path (Figure 2.6). The latter is very important for radiative transfer calculations (see Chapters 3 and 4). The spatial resolution of the instruments using this line-of-sight is approximately 1 km for imaging radiometers such as MODIS, and about 12 km for instruments such as IASI or GOSAT, whose high spectral resolution requires a much greater flow of electromagnetic energy. However, this type of sounder only measures one signal corresponding to the total column integrated from the ground to the satellite. Thus, the vertical distribution of the measurement is dependent on the technical characteristics of the instrument (spectral resolution, signal-to-noise ratio), and observation conditions (concentrations, thermal contrast). The vertical resolution, that is to say, the sharpness of the vertical distribution of the information contained in the measurement, can be estimated by using an information content analysis (see Chapter 5).

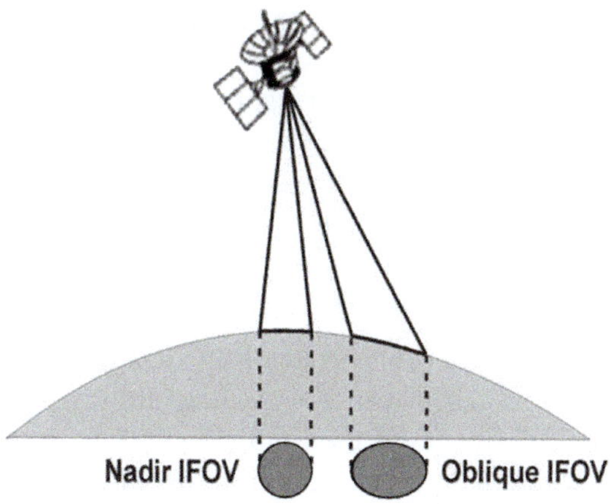

Figure 2.6. *Altering the IFOV and spatial resolution (horizontal) of a satellite in the nadir viewing with the observation angle. For example, for the instrument IASI, the spatial resolution is a circle with a diameter of 12 km at the nadir, at the end of the swath it is an ellipse with large axis of 39 km and short axis of 20 km*

In general, the characteristics of this line-of-sight make it a great tool for studying the surface properties and the physico-chemical properties of the troposphere (up to 20 km). They can also be used for weather or the air quality together with microwave or visible UV instruments, especially to access information in the boundary layer or on cloudy conditions.

2.1.3.1.1. Instruments observing at the Nadir

A number of infrared instruments use the Nadir line-of-sight including IMG, AIRS, MODIS, TES, IASI, or even TANSO-FTS (see Table A.2 of the Appendix).

For example, the IASI instrument [CLE 07], developed by CNES and EUMETSAT, is aboard the MetOp platform since 2006. It records the infrared emission spectrum at the Nadir line-of-sight (98,7° inclination) with an altitude of about 800 km and an IFOV (15.3 mrad), sampled by a 2x2 matrix corresponding to the circular pixels each with a diameter of 12 km. Measurements are taken every 50 km at the nadir with excellent spatial coverage (99% global coverage) due to its scanning capacity of 2200 km. It has already detected more than 25 different molecular species including measurements of total columns and the low vertical resolution profiles of ozone, carbon monoxide, water vapor and even methane, at both the regional and global scale. In addition, observations at the nadir that use infrared are possible both day and night since the source of radiation is independent of the Sun.

2.1.3.2. *Limb and solar occultation*

The limb line-of-sight (outer edge of the disk of a star), is to detect the light emitted/absorbed/scattered by the atmosphere at different altitudes. Figure 2.7 illustrates the limb viewing in the case of Earth's atmosphere modeled by different concentric circular layers corresponding to different altitudes. One of the techniques in the limb line-of-sight is the occultation mode, which involves observing the rising and lying of a celestial body, such as the Sun or Moon. The occultation measurements can be taken just as well in the visible UV range as in the infrared, depending on the components one wishes to study.

Successive measurements allow the altitude of clouds or aerosol plumes to be readily obtained, as well as the vertical profiles of gas concentrations with a vertical resolution defined by the instrumental characteristics, typically 1 to 3 km. Another advantage of the occultation mode is related to the optical path length of several hundred kilometers, which can detect certain molecules in very low concentrations. However, this line-of-sight has a limited horizontal resolution, related to the length of the optical path in the atmosphere (Figure 2.7). In addition, access to the troposphere below an altitude of 6 km is limited by the significant absorption of water vapor and clouds.

In solar occultation, the advantage is benefitting from a powerful and relatively stable source or radiation. However, the geographical coverage and sampling are limited by the number of sunrises/sunsets that can be observed, i.e. approximately 15 per day for instruments in low circular orbit. Lunar and, in particular, stellar occultations help increase the number of measurements per orbit. However, these require the design of a more elaborate instrument, with many moving parts, which can affect its stability and longevity in the space environment.

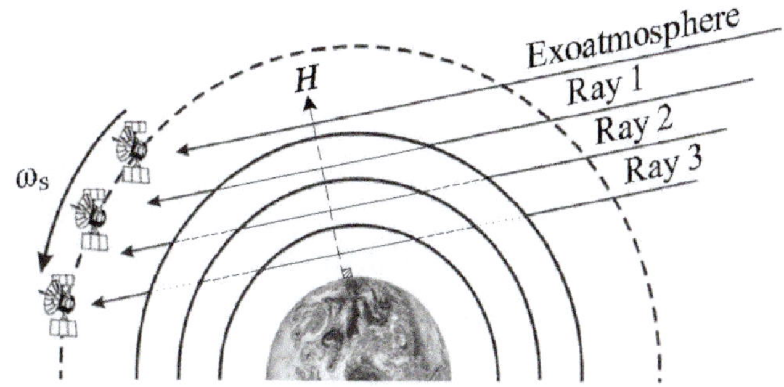

Figure 2.7. *Observation of the Earth's atmosphere from the limb viewing. Rays 1, 2 and 3 represent the observation of a star (e.g. the Sun) through atmospheric layers corresponding to different tangent altitudes H. The ray called "exoatmosphere" corresponds to the observation of radiation from a star (source) beyond any effects of its interaction with the Earth's atmosphere*

This line-of-sight is suitable for physical-chemical studies of the middle troposphere to upper stratosphere (5–60 km). However, it is not as suitable for weather or the air quality, and more generally for all applications that require a good sensitivity for measurements in lower atmospheric layers (<5 km).

2.1.3.2.1. Instruments observing at the limb

A number of instruments use the limb line-of-sight, including SCIAMACHY, CLAES, ILAS, MLS, ACE-FTS, OSIRIS, GOMOS, HALOE, SAGE, MIPAS, or even HIRDLS (see Table A.2 in the Appendix).

In particular, the Canadian instrument ACE-FTS [BER 05], exclusively uses the solar occultation technique. This has been in orbit onboard the SCISAT-1 platform since 2004 and is a Fourier transform spectrometer operating in the infrared range (500–5000 cm^{-1}) with a spectral resolution of 0.02 cm^{-1}. Its IFOV of 1.25 mrad allows it to attain a vertical resolution of approximately 1 km. Despite the low number of daily occultations available ($\leq$15), more than 70 gaseous components, as a trace or with a short lifetime, have been detected for the first time [COH 07].

2.2. Infrared detection techniques

In this section, we are specifically interested in techniques that exploit infrared, which is the favored spectral range of vibrational molecular transitions (see Chapter 3). With regard to observations of the Earth's surface, aerosols and hydrometeors, the electromagnetic radiation measured varies slightly with the wavelength (larger bandwidth). However, in the case of gaseous species, radiation consists of ro-vibrational transitions whose spectral variation is very large (high frequency) and their observation requires the use of instrumentation to separate the different wavelengths that make up the radiation. Thus, here we differentiate hyperspectral and radiometric instruments.

2.2.1. *Radiometers*

A radiometer is a receptor that measures the flow of electromagnetic energy. It consists of one (or more) sensor(s) and an optical part. The latter, which consists of a combination of lenses and mirrors, collects the radiation and transfers it in the plane of the detectors. In addition, a filter wheel can be used to select (or improve) the spectral bands of observation, each corresponding to a "channel". Thus, we can distinguish between imaging radiometers and spectro-radiometers according to the number and width of the spectral bands (Figure 2.8):

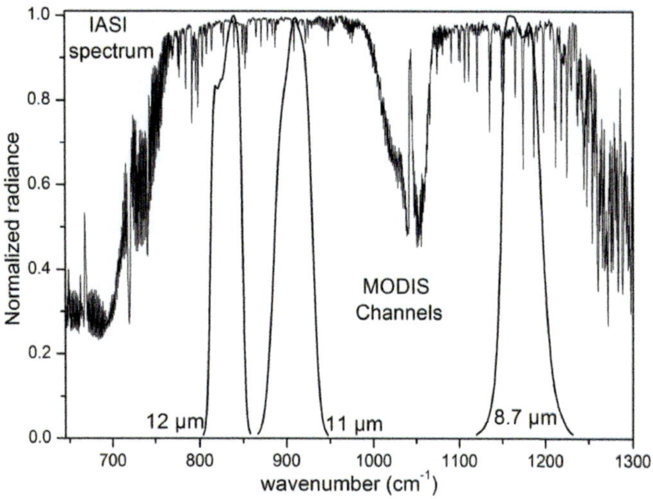

Figure 2.8. *Illustration of spectral channels at 8.7, 11 and 12 µm of MODIS, compared with a high spectral resolution lines spectrum from the IASI instrument between 645 and 1300 cm^{-1}*

– for imaging radiometers, the number of channels is low (e.g. three channels for satellites Météosat 1 to 7) and the bands are spectrally extended (wide). In particular, they aim to translate radiation measurements into images;

– spectro-radiometers have a large number of channels, and hence finely analyze the distribution of radiation according to wavelength. For example, the sounder ATOVS on NOAA satellites has 19 channels in the infrared range;

Sensors assure the conversion of the electromagnetic energy collected into electrical energy. There are two types of infrared detectors:

– thermosensors: these are sensitive to changes in temperature through a change in their physical properties: electrical resistance (bolometer), surface charge-capacity (pyrometer), or thermal expansion (Golay cell). Thermosensors can operate at room temperature and their detection threshold is independent of wavelength;

– photodetectors or quantum detectors: these absorb light energy, which generates charge carriers (photovoltaic or photoconductor effect) and excite the electrons in the conduction band, creating a current through the semiconductor material, proportional to the radiation intensity. They have the advantage of being miniaturized and used in matrices (CCD) for imaging. This is the case of the instrument MERIS. The main infrared photodetectors are photodiodes (HgCdTe and InSb), QWIPs (Quantum Well Infrared Photodetectors) and QDIPs (Quantum Dot Infrared Photodetectors). Photodetectors are faster and more sensitive, but the detection limit depends on wavelength and they usually have to be cooled to very low temperatures to avoid thermal noise.

In remote sensing, radiometers are most often used to determine the optical properties of clouds and aerosol plumes as well as surface properties (albedo, emissivity). However, radiometry can be coupled with correlation spectroscopy devices (cell correlation radiometry) to specifically examine a gaseous species. This technique is exploited by the instruments ISAMS, MAPS, MOPITT and SAMS, to determine the concentrations of trace and/or greenhouse gases.

2.2.2. *High spectral resolution instruments*

The terminology "high spectral resolution (HSR)" is commonly used in spectroscopy to identify the measurements taken with a spectral resolution close to the Doppler line-width (see Chapter 3). As this is dependent on the pressure and temperature of the environment, on the molecular species measured, and the characteristics of the instrument, the HSR term cannot be strictly applied. In general, especially in remote sensing, we call any type of measurement, HSR or hyperspectral, that can solve the ro-vibrational structure of gaseous molecular species. It is possible to directly measure spectra in the frequency space with spectro-radiometry if the frequencies are relatively low (λ>2000 μm or $\bar{\upsilon}$<5 cm^{-1}), beyond this, i.e. from the infrared, the frequencies are too high and their observation requires the use of a dispersive element. In general, instruments dedicated to HSR measurements are based on the principle of interferences of electromagnetic waves. In optics, many devices can therefore be used to disperse light in order to separate polychromatic radiation into different frequencies, such as Fabry–Perot or prism spectrometers. However, in remote sensing, the most common instruments are grating spectrometers and Fourier spectrometers. Both instruments are typical in infrared spectroscopy and here we will provide only a short summary.

2.2.2.1. *Grating spectrometers*

Spectra are etched onto reflective or transparent surfaces, or photographed on aluminized photosensitive resins (holographic grating). Although the diffraction of light by each grating line plays an important role by radiating in all directions, it is the interference phenomena between waves diffracted by many grating lines that cause the dispersion effect. The fundamental law of gratings is in reality a law of interference. It is also what gives the grating its directional selectivity. Although considered weakly bright, gratings are widely used as dispersers from infrared to ultraviolet. If we consider a planar array of parallel and equidistant straight lines, the essential parameter is step d (Figure 2.9(a)). When a monochromatic beam of parallel rays falls on such a grating with an angle of incidence α, light is observed in specific directions β, such that:

$$d\,(\sin\alpha \pm \sin\beta\,) = m\lambda \qquad [2.1]$$

where the signs – and + respectively concern transmission and reflection gratings, m being the integer that represents the grating interference order.

Diffraction by a grating slit determines the different directions in which light is directed, with each grating slit acting as a secondary source. The interference between these secondary sources construct the interference fringes, which are much finer when the grating has many lines (Figure 2.9(b)).

The grating then acts as a multiple-beam interferometer and the number of interfering waves is equal to the number N of lines.

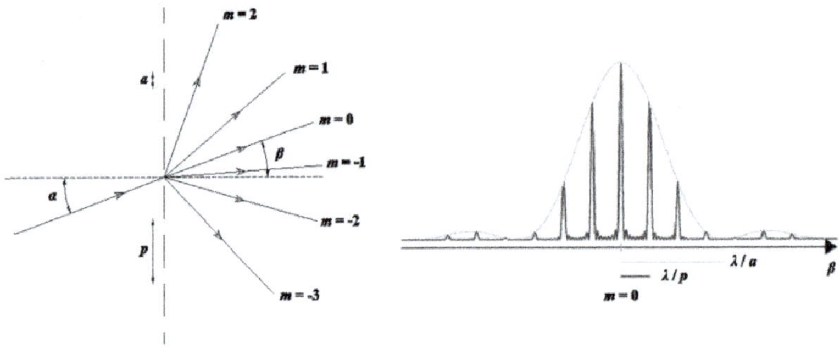

Figure 2.9. *a) Illustration of the orders of grating interferences for a monochromatic wave (left). b) Evolution of the interference signal as a function of the angle of the diffracted beam (right). Interferences (black) are formed mainly in the main lobe of the diffraction task (gray). The width of the interference peak is proportional to 1/N*

For a given value of m and α, β increases with wavelength. Thus, the grating diverges more red radiation than blue radiation.

Also, note that at $m = 0$ of the grating ($\alpha = -\beta$) corresponds to the direction of direct transmission for a transmission grating, and to the direction of specular reflection for a reflection grating.

By deriving expression [2.1], with respect to λ, and with α being constant, we obtain the angular dispersion produced by the grating:

$$D_\beta = \frac{\partial \beta}{\partial \lambda} = \frac{m}{d \cos\beta} \quad [2.2]$$

Note that the grating dispersion increases as one approaches the grazing incidence ($\alpha = 90^\circ$, $\cos\ \beta = 0$).

The spectral resolution of a grating spectrometer is determined by its resolving power:

$$\mathrm{R} = \frac{\lambda}{\delta\lambda} = \frac{mL}{d} = mN \quad [2.3]$$

with L the grating width, N the total number of lines such that $N = L/d$.

From these expressions [2.2] and [2.3], it is clear that to achieve high dispersion and a high resolving power, the highest order possible must be used. However, the transmission grating has the disadvantage of allowing light in order 0 to pass and the greater the order is, the poorer the overlay of adjacent orders. In this case, a way to increase the brightness on particular orders is to use echelette or "blazed" gratings (Figure 2.10). These are blazed gratings used in reflection, for which each grating line corresponds to a step and acts as a small mirror. If the order *m* for a wavelength λ_0 is centered on the maximum of the diffraction figure of one of these mirrors, all other orders will coincide with 0 in this diffraction figure. The result is that all the light is diffracted in this order *m*, allowing the grating to achieve a very high efficiency. However, these systems are effective if *d* of the grating is larger than the wavelength λ_0 of the radiation studied. Now, if for a width *L* of a given grating, *d* is large, the total number of lines *N* is small. Thus, given the technical challenges of implementing such a grating with full control of its lines, its use is still restricted to wavelengths that are not too high or infrared measurements that do not require too much resolving power. In case where high spectral resolution measurements are required at long wavelengths, planar arrays are preferred.

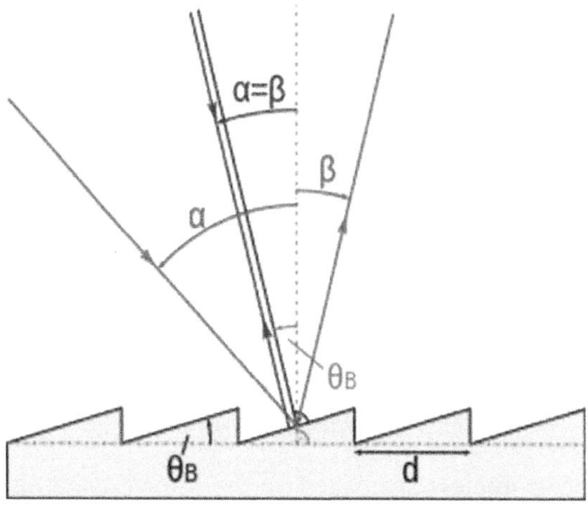

Figure 2.10. *Blazed grating. α is the angle of incidence, β is the angle of the diffracted beam and d is the grating step*

In practice, grating spectrometers used in remote sensing such as AIRS or OCO, operate in reflection and consist of an optical feature that eliminates chromatic aberrations and unwanted reflections. The operating principle is as follows (Figure 2.11): a ray of light coming from an entrance slit enters a concave mirror that reflects and focuses the light on the whole grating. Each line of the grating guides the light according to the wavelength with an angular dispersion defined by equation [2.2]. A second mirror may be used to focus the radiation onto the array detector with a wavelength range defined by the resolving power of the grating (see equation [2.3]). Generally, the latter is formed of a photodiode array.

In remote sensing, this type of instrumentation has a number of advantages, such as small footprint, low-cost energy, low mechanical stress and low sensitivity to pointing.

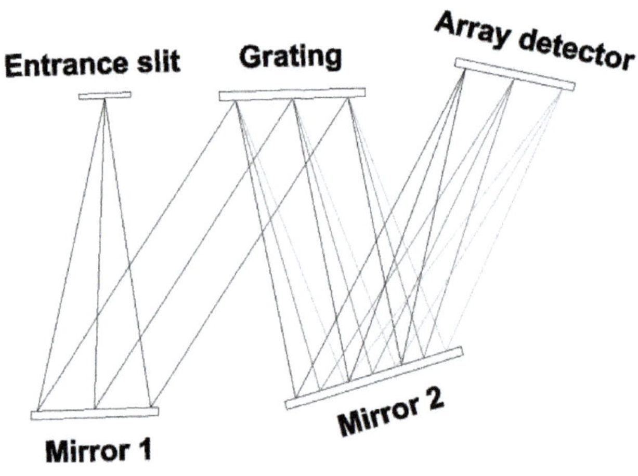

Figure 2.11. *Principle of grating spectrometer in reflection*

2.2.2.2. *Fourier transform spectrometer*

Fourier spectrometers are based on the interference phenomenon at two electromagnetic waves. They consist of recording a spectrum in the space of distances using an interferometer, the most common being a Michelson interferometer (Figure 2.12). The latter is composed of a semi-reflecting beam-splitter S, a fixed mirror M_1 and a mobile mirror M_2 respectively located at distances d_1 and d_2 of S. A ray of light arrives at S, which reflects some of the radiation toward the fixed mirror M_1 and transmits the other part to the moving mirror M_2. Both beams are then reflected by the mirrors M_1 and M_2 and are sent to the detector via S. In practice, a compensating plate C, is added between S and M_2, to ensure that the two rays arriving at the detector have passed through the same plate thickness. Both beams reach the detector, having passed through a different optical path, with a path difference Δ proportional to the difference $(d_1\text{-}d_2)$. The latter therefore depends on the position of the mobile mirror. In the case of a monochromatic wave, the light intensity is written: $I = I_0 [1+\cos(2\pi\nu_0\Delta)]$. Thus, the waves will interfere constructively for integer values of $\nu_0\Delta$ and destructively (no signal) with half-integer values. At a path difference of zero (i.e. Δ=0) the signal is at its maximum, since it corresponds to the case where all waves are in phase. The signal of a polychromatic source will result from the sum of the signals of each monochromatic component constituting the source. This sum I is the interferogram (Figure 2.13):

$$I(\Delta) = \int_0^{\infty} B(\nu)[1 + \cos(2\pi\nu\Delta)]\, d\nu \qquad [2.4]$$

where $B(\nu)$ is the spectral energy density characterizing the source, and ν the wavenumber.

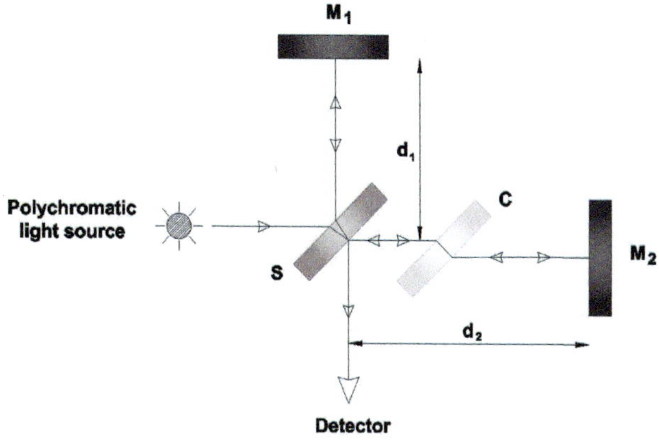

Figure 2.12. *Principle of a Michelson interferometer*

Equation [2.4] consists of a constant term and a modulated even term, that is to say, dependent on the path difference, which is the real part of the Fourier transform in cosine of $B(\nu)$. The inverse Fourier transform of the interferogram:

$$B(\nu) = \int_{-\infty}^{\infty} I'(\nu)\cos(2\pi\nu\Delta)\, d\Delta, \qquad [2.5]$$

allows us to obtain the spectrum in the frequencies space.

Thus, the interferogram is firstly recorded as a function of Δ, then its Fourier transform returns the spectrum as a function of ν.

However, equation [2.4] assumes an infinite extension movement of the mobile mirror, when in practice, the optical path difference is limited to a maximum value Δ_{max}.

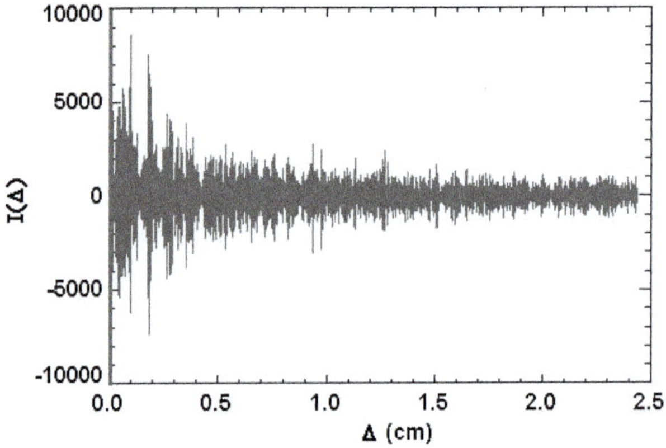

Figure 2.13. *Example of an interferogram corresponding to a maximum path difference of 2.5 cm*

We therefore do not measure the function $I(\Delta)$, but its product by a rectangular function which is equal to zero beyond Δ_{max} (Figure 2.14(a)). Thus the frequency spectrum is convoluted by a cardinal sine (sinc) function f expressed as:

$$f = \sin(2\pi\nu_0\Delta_{max})/(2\pi\nu_0\Delta_{max}) \qquad [2.6]$$

This convolution induces oscillations around the central line (Figure 2.14(b)), whose Full Width at Half Maximum (FWHM), imposed by Δ_{max}, gives the spectral resolution limit δ_v:

$$\delta_v = 1/(2\,\Delta_{max}) \qquad [2.7]$$

To decrease the large secondary maxima of the cardinal sine function, we can multiply the interferogram using an apodization function, which can take different forms (box-car, triangular). This can significantly reduce the intensity of oscillations, but it does, however, degrade the spectral resolution.

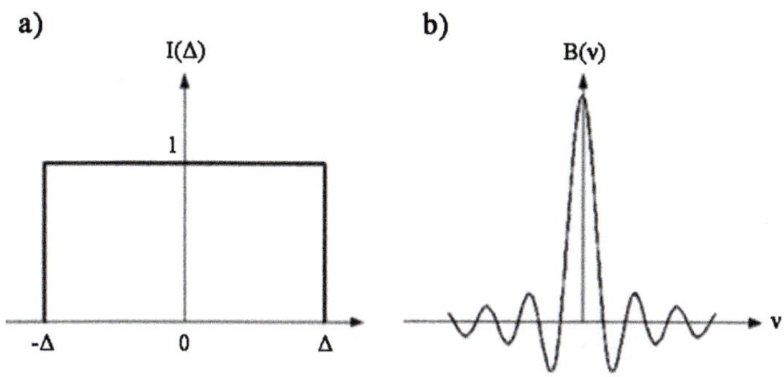

Figure 2.14. *a) Rectangular function used for the integration of the interferogram over a finite distance field (apodization). b) Illustration of the oscillations in the spectrum by apodization*

This technique has some advantages, compared to grating spectrometers for three main reasons:

1) *The multiplex advantage (or Felgett's advantage)*

The multiplex approach allows a broad spectrum to be analyzed with a single detector. The interferogram is in fact the sum of signals modulated proportionally to v.

This approach, unlike sequential recording, improves the signal-to-noise ratio of a spectrum consisting of M spectral components by $(M)^{1/2}$. By comparison, a grating spectrometer detects at every moment, only a narrow band of radiation. This multiplex advantage has become less used with the development of increasingly powerful detectors. That being said, with all other things being equal, the range spanned by the source observed is about 100 times greater than that reached with a grating spectrometer. Fourier transform spectrometry will therefore be much faster than a grating spectrometer.

2) *The throughput advantage (or Jacquinot's advantage)*

The narrow input slit of a grating spectrometer, which severely limits the amount of transmitted radiation, is replaced by a circular opening with a large surface area. This is advantageous in that a large proportion of the radiation source is used for measuring, which results in a large signal-to-noise ratio.

3) *The accuracy advantage (Connes' advantage)*

The instrumental distortion effects that affect the quality of interfering wave surfaces remain the same during a recording. This leads to a perfect understanding of the apparatus function. In addition, calibration using monochromatic radiation (e.g. HeNe laser) accurately fixes the scale of wavenumbers.

All these advantages justify that this instrumental device is currently the most widespread in high spectral resolution spectrometry, either for laboratory measurements or satellite remote sensing.

2.3. For further information

Satellite and spatial remote sensing

[MAI 11] MAINI A.K., Agrawal V., *Satellite Technology: Principles and Applications*, Wiley and Sons Publications, 2011.

[NRC 08] NATIONAL RESEARCH COUNCIL, *Earth Observations from Space: The First 50 Years of Scientific Achievements*, Committee on Scientific Accomplishments of Earth Observations from Space, Board on Atmospheric Sciences and Climate, Division on Earth and Life Studies, The National Academies Press, 2008.

[PUR 11] PURKIS S.J., KLEMAS V.V., *Remote Sensing and Global Environmental Change*, Wiley-Blackwell, 2011.

Fourier transform spectrometry: theory and applications

[DAV 01] DAVIS S., ABRAMS M., BRAULT J., *Fourier Transform Spectrometry*, Academic Press, 2001.

[GIF 07] GIFFITH P.R., DE HASETH J.A., *Fourier Transform Infrared Spectrometry*, John Wiley & Sons, 2007.

Infrared detectors

[KIN 07] KINCH M.A., *Fundamentals of Infrared Detector Materials*, SPIE, 2007.

[ROG 10] ROGALSKI A., *Infrared Detectors*, 2nd ed., CRC Press Taylor and Francis group, 2010.

3

Forward Radiative Transfer in Absorbing Atmosphere

To understand why passive remote sensing techniques, particularly optical infrared measurements, are powerful tools for analyzing the atmosphere, it is essential to understand the physics that govern the radiation–matter interaction. Thus, this chapter is dedicated to establishing the link between the fundamental properties of molecules and their manifestations at the macroscopic scale, induced by infrared radiation. Indeed, just like digital fingerprints that are unique to each individual, how the material reacts to radiation is specific to each component. Thus, understanding the process of absorption and emission makes it possible to specifically detect gaseous species, and in the best of cases to quantify their concentration in the atmosphere. In this chapter, we will discuss the processes of absorption and emission of radiation in the gaseous phase. We will then define all the parameters necessary for the formulation of the radiative transfer equation in a purely absorbing atmosphere, which can account for the interaction between telluric and/or solar radiation and gaseous components that make up the Earth's atmosphere.

3.1. Gaseous absorption and emission

3.1.1. *Overview*

If we consider the process of absorption/emission between the energy levels of a molecule, the corresponding transition is such that:

$$\Delta E = E_k - E_l = h\nu = hc/\lambda = hc\tilde{\nu} \qquad [3.1]$$

where E_k and E_l are, respectively, the energies of states l and k, h is Planck's constant, c the speed of light in a vacuum, ν the frequency, λ is the wavelength, and $\tilde{\nu}$ is the wavenumber.

All the transitions for the same molecule form an energy spectrum, which can be determined by solving Schrödinger's equation:

$$\hat{\mathrm{H}}\psi_n(\boldsymbol{r},\mathrm{t}) = E_n\psi_n(\boldsymbol{r},\mathrm{t}) \qquad [3.2]$$

where $\psi_n(\boldsymbol{r},\mathrm{t})$is the characteristic wave function of the quantum state$n$, such that: $\psi_n(\boldsymbol{r},\mathrm{t}) = \psi_n(\boldsymbol{r})\exp\left(-i\frac{E_n t}{h}\right)$, E_n is the energy of state n, and $\hat{\mathrm{H}}$ is the Hamiltonian operator. The latter is the sum of the kinetic energy T of the nuclei and electrons and the potential energy V, which comprises V_{ee} and V_{nn} due to the electron–electron and nuclei–nuclei Coulombic repulsion as well as a third term V_{en}, due to the forces of attraction between the electrons and the nuclei. If we consider a molecule composed of a nucleus with mass M_N and i electrons with mass m_e, the origin of the coordinate system being the molecule's center of mass, the internuclear distance is denoted $\boldsymbol{R}_N$ and the position of the electrons is given by the vectors $\boldsymbol{r}_i$, the Hamiltonian stationary (i.e. independent of time) of the molecule at the non-relativist approximation is written:

$$\mathrm{H} = T_n + T_e + V_{nn}(\boldsymbol{R}_N) + V_{ee}(\boldsymbol{r}_i) + V_{en}(\boldsymbol{r}_i, \boldsymbol{R}_N) \qquad [3.3]$$

with, $T_n = -\frac{\hbar^2}{2\mu}\overrightarrow{\nabla_R}^2$ and $T_e = -\frac{\hbar^2}{2m_e}\overrightarrow{\nabla_{r_i}}^2$

As things stand, exactly solving equation [3.3] is impossible for molecular systems. However, if one considers that the mass of nuclei is much greater than that of the electrons, it appears that the movement of the nuclei is very slow compared to the movement of the electrons. This instantaneous adaptation of electrons to any new nuclear configuration is the foundation of the Born–Oppenheimer approximation (also known as the adiabatic approximation), which states that one can separate the electrons and nuclei variables into two distinct groups: the electron and nuclear movements are decoupled, and within the nuclear movement the rotation and vibration are also such that the energy in the equation [3.2] can be written as:

$$E_n = E_n^{elec.} + E_n^{vib.} + E_n^{rot.} \qquad [3.4]$$

where the terms $E_n^{elec.}$, $E_n^{vib.}$ and $E_n^{rot.}$ correspond respectively to the electronic, vibrational and rotational energies. The energy associated with the change in electron distribution is approximately 10^4–10^6 cm^{-1} and can therefore be observed in the visible-UV range of the electromagnetic spectrum. The energy due to nuclei vibrations is in the infrared between 10^2 and 10^4 cm^{-1}, whereas that associated with rotations is approximately 1 to 100 cm^{-1}, corresponding to micro-waves and far-IR.

The spectral range of interest here is between 500 and 13000 cm^{-1}, in which spectra of rovibrational transitions can be observed.

The probability that a transition between energy levels takes place according to equation [3.1] is called the transition probability or force of transition. This probability is determined as being the square of the moment of transition $\mathcal{M}^2$. For the electron dipole approximation, at a given wavelength, the moment of transition is written:

$$\mathcal{M}_{k,l} = \langle\psi_k|\boldsymbol{D}|\psi_l\rangle = \int \psi_k^* q.\boldsymbol{r}\psi_l \qquad [3.5]$$

where $\boldsymbol{D}$ is the electron dipole moment, and q is the electron charge whose position indicated by vector $\boldsymbol{r}$.

So, when the integral [3.5] is zero, the transition between states k and l will be "forbidden". Otherwise, especially for molecules with complex symmetry that have dipole moments ($\mathcal{M}_{k,l} \neq 0$), the processes of absorption/emission can occur *via* the bias of "allowed" transitions. However, there are some restrictions, known as selection rules.

In what follows, we will endeavor to concisely explain how to establish distribution levels of vibrational and rotational energy as well as the selection rules governing the transitions between these energy levels from simple theories of Physics, usually from classical mechanics.

3.1.2. *Rovibrational spectroscopy*

The infrared range of the electromagnetic spectrum consists mainly of vibrational transitions between different levels of the same electronic state, with a structure composed of transitions between different rotational levels.

3.1.2.1. *Vibration of diatomic molecules*

In the case where diatomic molecules consist of two atoms (masses m_1 and m_2), the vibrational motions can be modeled using classical mechanics, with the help of the harmonic oscillator approximation, which considers that atoms connected by a spring obeying Hooke's law:

$$f = -\frac{dV(x)}{dx} = -kx \qquad [3.6]$$

where V is the energy potential, k the rigidity constant and x is the movement with regard to equilibrium (elongation or compression).

The fundamental relation of the dynamics is therefore written as: $\mu \frac{d^2x}{dt^2} + kx = 0$, where μ is the reduced mass, such that: $\frac{1}{\mu} = \frac{1}{m_1} + \frac{1}{m_2}$. The solution shows that the movement oscillates sinusoidally (or harmonically) with an eigen pulsation: $\omega = \left(\frac{k}{\mu}\right)^{1/2}$.

Integrating equation [3.6] gives: $V(x)=1/2\ kx^2 + C$, with C a constant. This equation corresponds to that of a parabola; the Hamiltonian parabola is given by:

$$H=-\frac{\hbar^2}{2\mu}\frac{\partial^2}{\partial x^2}+\frac{1}{2}kx^2 \qquad [3.7]$$

Equation [3.2] therefore becomes:

$$\frac{\partial^2\psi}{\partial x^2}+\frac{2\mu}{\hbar^2}\left[E_n^{vib.}-\frac{1}{2}kx^2\right]\psi(x)=0 \qquad [3.8]$$

The solutions to the previous equation correspond to the energies limited to the following quantified values:

$$E_n^{vib}(v)=\hbar\omega\left(v+\frac{1}{2}\right)=\omega_e\left(v+\frac{1}{2}\right), v\in\mathbb{N} \qquad [3.9]$$

where v is the quantum number of vibration and ω_e is the vibrational constant in cm^{-1}.

The vibrational energy levels are therefore equidistant from $\hbar\omega$. This regular spacing between the energy levels is a structural property of the harmonic oscillator. However, this is only valid if x is small, that is to say that $\boldsymbol{R}_N$ is close to the equilibrium distance. When $\boldsymbol{R}_N$ becomes large, the molecule dissociates: two neutral atoms are formed and, since they do not have an influence on each other, the force constant is zero and $\boldsymbol{R}_N$ can increase without the potential V changing. Therefore, the potential must "flatten" to $V = D_e$ where D_e is the dissociation energy measured relative to the minimum potential (Figure 3.1). Moreover, on approaching the dissociation threshold, the force constant tends to zero and the bonds weaken. This causes the potential energy curve to become less steep than for the harmonic oscillator. At low $\boldsymbol{R}_N$ values, the positive charges of both nuclei repel each other, which prevents them from becoming closer. Consequently, the curve is steeper at short distances than for the harmonic oscillator.

The deviations from the harmonic potential curve that we have just described are due to mechanical anharmonicity.

Anharmonicity, can be analyzed using Morse's potential energy in the form (Figure 3.1):

$$V(x) = D_e(1 - e^{-ax})^2 + V(r_e). \qquad [3.10]$$

where a is a characteristic constant of the electronic state of the molecule. This new expression of potential energy taking into account the anharmonicity modifies the vibrational term of the energy in equation [3.9], such that:

$$E_n^{vib.}(v) = \omega_e\left[(v+\frac{1}{2}) - x_e(v+\frac{1}{2})^2\right] \qquad [3.11]$$

where x_e is the anharmonicity constant. The sign of $\omega_e x_e$ [cm^{-1}] is always positive, which brings the vibrational levels increasingly closer as v increases, (see Figure 3.1), which converge at the dissociation limit $D_e \approx \frac{\omega_e^2}{4\omega_e x_e}$, beyond which there is a continuum of levels.

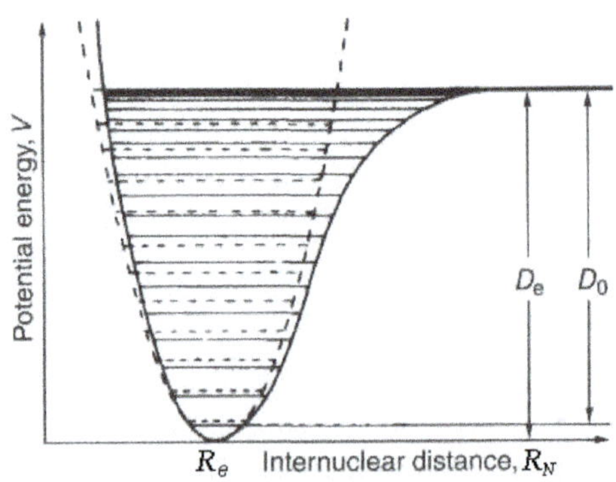

Figure 3.1. *Evolution of potential energy with internuclear distance R_N. The dotted lines represent the harmonic oscillator. The straight lines represent Morse's potential energy. D_e is the dissociation energy of the molecule, R_e is the distance from the nuclei at equilibrium and D_0 represents the energy difference at the dissociation limit and the energy of the first vibrational state v=0*

3.1.2.2. *Vibration of polyatomic molecules*

Spectra of molecules formed from more than two atoms are more complicated, because their movements due to vibrations are more numerous. Therefore, a molecule formed from N atoms is a set of N coupled oscillators. To use or interpret polyatomic spectra, it is necessary to consider the shape of the molecule, i.e. the internal arrangement of atoms and study their properties of symmetry are described by group theory in terms of symmetry point groups [HER 66].

We can visualize a polyatomic molecule from a certain number of point masses united by springs. If such a system receives energy, it will carry out a complicated vibrational movement (stretching, deformation, swinging, wagging, twisting) an example being a molecule of CO_2 in Figure 3.2. The decomposition of such a movement into simpler movements can be obtained using the table of characters of the molecular symmetry point group. This helps decompose a complicated movement of a system of material points into a series of simpler movements by assuming that at each instant, the position of the point system will be given by the geometric sum of positions that the system would have if it were only subject to elementary movements. These elementary movements (irreducible representation of a molecular symmetry point group) are known as normal modes of vibration. These are independent of each other, and for each of them, the nuclei vibrate in phase and with the same frequency.

The number of normal modes of vibration can be determined following simple considerations: each of the N atoms of a nonlinear polyatomic molecule has three degrees of freedom. Of the $3N$ degrees of freedom of the molecule, three correspond to a translation along the three axes and three correspond to the rotation of the molecule about each axis. This leaves $3N$ - 6 degrees of vibrational freedom. A linear molecule has $3N$ - 5 normal modes since it only has two degrees of rotational freedom.

For these normal modes, the vibrational energy terms are identical to the equation of diatomic molecules by replacing the reduced mass by a fictitious mass of the mode.

The most general vibratory movement of a molecule results in superposition (linear combination), with suitable amplitudes and phases, of these $3N - 6$ (or $3N - 5$) fundamental vibrations.

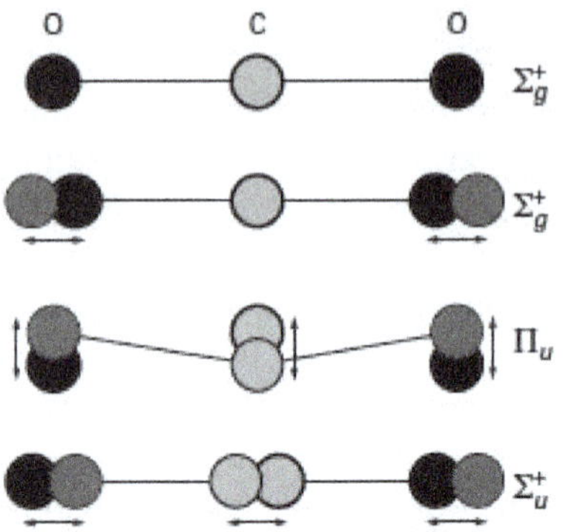

Figure 3.2. *Fundamental vibrations (normal modes of vibration) of the molecule of carbon dioxide*

3.1.2.3. *Vibrational spectrum*

A heteronuclear diatomic molecule has a nonzero dipole moment which varies with the x stretching, the latter remaining small. The change in the electric dipole moment $\boldsymbol{D}$ can therefore be expressed using a series expansion:

$$D = D_e + \left(\frac{\partial D}{\partial R_N} \right)_e R_N + \frac{1}{2!} \left(\frac{\partial^2 D}{\partial R_N{}^2} \right)_e R_N{}^2 \qquad [3.12]$$

where D_e is the electric dipole moment at equilibrium.

The moment of transition is therefore:

$$\mathcal{M}_{k,l} = D_e \left\langle \varphi_{vk,J'} \middle\| \varphi_{vl,J} \right\rangle + \left(\frac{\partial D}{\partial R_N} \right)_e \left\langle \varphi_{vk,J'} \middle| R_N \middle| \varphi_{vl,J} \right\rangle \qquad [3.13]$$

The first term in the right side of the equation is zero when the wave functions are orthogonal. The dipole moment of a homonuclear diatomic or polyatomic molecule, which has a center of inversion (i.e.

unvarying symmetry during the vibration), is zero. It is therefore not possible to observe a vibration spectrum. This is the case for the most abundant gases in the atmosphere, O_2 and N_2, which are linear symmetrically charged diatomic molecules. This explains why the latter, despite their high concentrations in the atmosphere, do not exhibit vibrational transitions in the infrared range of the spectrum. The second term is nonzero when $\Delta v = \pm 1, \pm 2, \pm 3, \ldots$ which establishes the selection rule for vibration. Therefore, all that is required is that there is a variation in the dipole moment during the vibration, i.e. $\frac{d\mu}{dr} \neq 0$ for a vibrational transition to occur. Therefore, all molecules (except homonuclear molecules) have a vibration spectrum. However, all vibrational modes are not necessarily active in the infrared and depend on the degree of symmetry.

3.1.2.4. *Rotation of diatomic molecules*

In the case of diatomic molecules (or polyatomic linear molecules) that rotate about their center of mass (Figure 3.3), the rotational energy can be typically described using the rigid-rotator model whose kinetic energy is equal to:

$$T = \frac{1}{2} I \omega^2 = \frac{L^2}{2I} \qquad [3.14]$$

where $I = \mu R_N{}^2$ is the moment of inertia, ω is the angular rotational velocity and $L = I\omega$ is the angular moment of rotation.

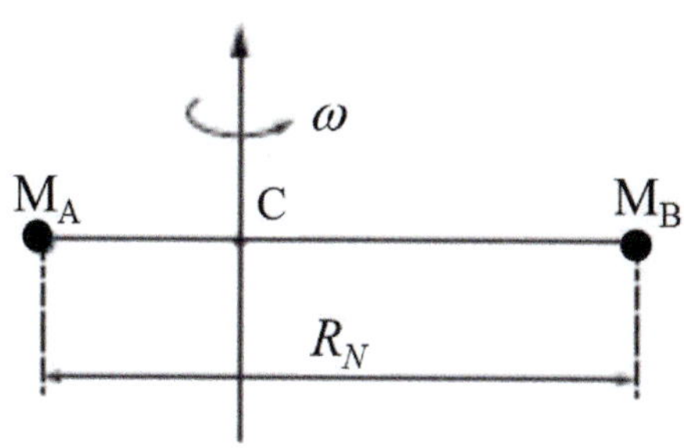

Figure 3.3. *Rotation of a diatomic molecule (M_A, M_B) with an angular velocity ω about an axis passing through the center of mass C*

There is no potential energy because, in the absence of any external force, the energy of the molecule does not depend on its orientation in space. The Hamiltonian operator is simply the kinetic energy operator. To determine the eigenvalues of the energy $E_n^{rot.}$ it is necessary to solve the angular part of equation [3.2], which is identical to that of a hydrogen atom, such that:

$$\frac{1}{2I}\hat{L}^2\,\psi_n(\theta,\varphi) = E_n^{rot.}\psi_n(\theta,\varphi) \quad [3.15]$$

The eigenfunctions of $\widehat{L^2}$ are spherical harmonics $Y_{lm}(\theta,\varphi)$, with l the quantum number of the orbital angular moment, and m the associated magnetic quantum number whose eigenvalues, are in the form $l(l+1)\,\hbar^2$ with $l \in \mathbb{N}$. With regard to molecules, we substitute l (atomic notation) for J (the total kinetic moment, $j = l + s$). The eigenfunctions $Y_{jmj}(\theta,\varphi)$ are such that:

$$\frac{1}{2I}\hat{L}^2\,Y_{J,M_J}(\theta,\varphi) = J(J+1)Y_{J,M_J}(\theta,\varphi) \quad [3.16]$$

whose associated eigenvalues define the following quantified rotational energy:

$$E_n^{rot.} = \frac{\hbar^2}{2I}J(J+1) = BJ(J+1)\; J \in \mathbb{N} \quad [3.17]$$

with the constant of rotation B [cm^{-1}]$= h^2/8\pi^2 I = h^2/8\pi^2\mu R_N^2$.

In addition, the component on the z axis (the axis perpendicular to the internuclear axis) of the spatial quantification of the angular moment of rotation is described by: $(L_j)_z = M_j\hbar$, where M_j varying from $-j$ to $+j$, i.e. $2j + 1$ eigenvalues. So, in the absence of an external field, each rotational level degenerates such that: $g_n = (2J+1)$.

The rigid-rotator model is very basic and it should actually represent the bond between atoms by a spring linking the nuclei. It is therefore not surprising, when the rotational velocity increases (J increases), that the nuclei tend to distance themselves from one another under the effect of centrifugal forces. As the spring extends, the interatomic distance increases and the constant of rotation B decreases. In fact, if we want to keep the constant of rotation as expressed above, it is necessary to add a negative corrective term:

$$E_n^{rot.} = BJ(J+1) - DJ^2(J+1)^2 \qquad [3.18]$$

D is the centrifugal distortion constant, which is always positive for a diatomic molecule.

3.1.2.5. *Rotation of polyatomic molecules*

A polyatomic molecule can rotate about three main axes with three different associated moments of inertia. The moment of inertia about an axis is defined as:

$$I = \sum_i m_i r_i^2 \qquad [3.19]$$

with m_i the atomic mass i and r_i the distance between the atom i and the axis. The axis origin is situated at the center of mass and the moments of inertia I_x, I_y and I_z can be used to classify molecules into four types:

1) $I_x=I_y$;I_z =0: diatomic or linear molecules (CO, C_2H_2)

2) $I_x=I_y=I_z$: spherical rotators (CH_4, CCl_3, SF_6)

3) $I_x=I_y \neq I_z$: symmetric rotators (BF_3, CH_3Cl, NH_3, C_2H_6)

4) $I_x \neq I_y \neq I_z$: asymmetric rotators (H_2O, CH_3OH)

However, conventionally, we denote c and a the axes for which I reaches its maximum and minimum value, respectively; the third axis perpendicular to the two others is denoted b. This helps classify molecules not only by axes of symmetry but also by size of moment of inertia. Therefore:

1) $I_a < I_b = I_c$: prolate symmetric rotators (or elongated) (BF_3)

2) $I_a = I_b < I_c$: oblate symmetric rotators (or flattened) (CH_3Cl)

We have seen above that the rigid-rotator energy is a function of the rotational parameters, which are inversely proportional to the moments of inertia:

$$A = \frac{\hbar^2}{2I_a} \quad B = \frac{\hbar^2}{2I_b} \quad C = \frac{\hbar^2}{2I_c}$$

In general, the Hamiltonian of rotation of a polyatomic molecule is written:

$$H = A J_A^2 + B J_B^2 + C J_C^2 \qquad [3.20]$$

Thus, for spherical molecules, the energy can simply be written, in the absence of anharmonicity, in the following form:

$$E_r = B J (J + 1) \text{, where} \quad J^2 = J_A^2 + J_B^2 + J_C^2$$

This is identical for linear molecules. However, as there are only two degrees of freedom to describe the rotation (about an axis perpendicular (*b* or *c*) to the molecule axis), then:

$$J^2 = J_B^2 + J_C^2.$$

With regard to symmetric molecules the rotational overall movement can be separated into the rotation about axis *a* of symmetry and a movement about an axis perpendicular to it. For an elongated molecule ($B = C$), the Hamiltonian is therefore written:

$$H = A J_A^2 + B\left(J_B^2 + J_C^2\right) = B J^2 + (A - B) J_A^2 \qquad [3.21]$$

The corresponding energy is therefore defined as:

$$E_r = B J (J + 1) + (A - B) K^2 \qquad [3.22]$$

where K is the quantum number corresponding to the projection of J on axis a.

If the molecule is flattened ($A = B$), we can easily show that the molecule can be expressed as follows:

$$E_r = B J (J + 1) + (C - B) K^2 \qquad [3.23]$$

Here, we are not concerned with asymmetric molecules, that is to say, with three different axes of symmetry, since it is difficult to obtain analytical expressions of their rotational energies.

3.1.2.6. *Rotational spectrum*

If we consider the pure rotation ($\Delta v = 0$), for a rotational transition to occur, it implies that the molecule has a permanent non-zero dipole moment. Spherical molecules do not normally have a pure rotation spectrum. However, in some cases, the centrifugal distortion can create an instantaneous dipole moment and consequently induce transitions. According to equation [3.5] in the rigid-rotator approximation, a rotational transition will be allowed if $\Delta J=\pm 1$ (with $\Delta K = 0$ for symmetric molecules), and the corresponding wavenumber will be in the form:

$$\tilde{\nu} = 2B\,(J+1) \quad [3.24]$$

So, for a purely rotational spectrum, the lines are equidistant and separated by $2B$ cm^{-1}.

If we consider a rotational transition when the diatomic molecule is in an excited vibrational state, we can show that the constant of rotation has a vibrational dependency according to:

$$B_v = B_e - \alpha (v + \frac{1}{2}) \quad [3.25]$$

where the index e is relative to the state of equilibrium, v is the vibrational level considered and α is the vibration–rotation interaction constant.

The wavenumber of rotational transitions in an excited state therefore becomes:

$$\tilde{\nu} = 2B_v(J+1) - 4D_v(J+1)^3, \quad [3.26]$$

with B_v and D_v which have a vibrational dependency such that:

$$B_v = B_e - \alpha_e\,(v+1/2) \text{ and } D_v = D_e + \beta_e\,(v+1/2)$$

Note that B_e helps experimentally determine the length of the inter-atomic bond at equilibrium, and D_e=$4B_e^3/\omega_e^2$, with ω_e the constant of vibration at equilibrium.

3.1.2.7. *Rovibration*

In rovibrational spectroscopy, transitions between energy levels of rotational energy are associated with different levels of vibration. However, it is important to note that if all vibrational transitions involve transitions between levels of rotation, the latter will only be observable in the gaseous phase and for not too heavy molecules. Otherwise, the rotational structure cannot be solved, which will be the case for liquids, solids, but also clorofluorocarbons (CFCs) (gases). In fact, they are often composed of many heavy atoms (i.e. large μ), which results in small B values, and therefore a difference in a short wavelength between each transition.

By applying the Born–Oppenheimer approximation, the total energy of a molecule is the sum of its electronic vibrational and rotational energies. If we consider the fundamental electronic state, which here we assume to be constant, the rotation-vibration energy is written:

$$E_n(v,J) = E_n^{vib.}(v) + E_n^{rot.}(J) = B_v J(J+1) - D_v J^2(J+1)^2 + \omega_e(v+1/2) - \omega_e x_e(+1/2)^2 \quad [3.27]$$

The selection rules defining the rovibrational transitions are: $\Delta v = \pm 1, \pm 2, \pm 3, \ldots$ and $\Delta J = \pm 1$.

In a spectrum of a diatomic molecule, there are two sets of lines, one at high frequencies, called "branch R" which is due to transitions for which $\Delta J = +1$ and the other set at low frequencies, called "branch P" due to transitions corresponding to $\Delta J = -1$ (Figure 3.4). For example, if we consider the transition $\Delta v = 0 \rightarrow 1$, in the absence of centrifugal distortion and anharmonicity of vibration, the energy of transition of branch R, is therefore:

$$h\nu_{obs} = E(1,J+1) - E(0,J) = h\nu + 2B_1 + (3B_1 - B_0)J + (B_1 - B_0)J^2 \quad [3.28]$$

Similarly for branch P, we obtain:

$$h\nu_{obs} = E(1, J-1) - E(0, J) = h\nu - (B_1 + B_0)J + (B_1 - B_0)J^2 \quad [3.29]$$

Like $B_1 < B_0$ due to the rotation vibration interaction and the fact that the length of the bond increases when v increases (equation [3.26]), the spacing between the lines in branch R decreases whereas in branch P it increases, as shown in Figure 3.5. This figure represents the transition $v'=1-v''=0$ of two isotopologues $H^{35}Cl$ and $H^{37}Cl$ (molecular species differing from one another simply by the presence of one or several isotopes). Note that the intensity ratio comes from natural abundance ratios of isotopes of chlorine and that the slight spectral shift is related to their mass difference (equations [3.9] and [3.17]).

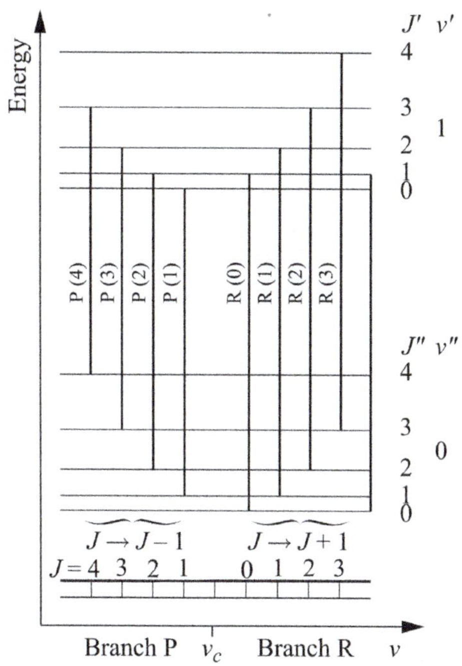

Figure 3.4. *Energy diagram of transitions of a rovibrational band. Note that their molecular transitions are traditionally written with the highest state first (unlike with atoms), the sign ' designates the highest state and the sign " indicates the lowest state*

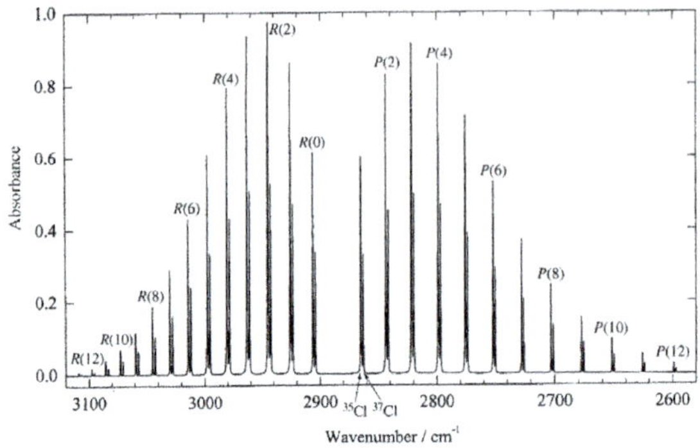

Figure 3.5. *Transition spectrum v'=1-v''=0 of molecules H^{35} Cl and H^{37} Cl. Both isotopes of chlorine are present in their natural abundances*

For some transitions, the couplings between vibration and rotation will slightly modify equation [3.2]. There also may be many couplings [LEF 86] of different natures, which can usually be analyzed using the perturbations method. The most commonly encountered effect is related to the geometry of molecules which can induce a moment of perpendicular transition at the internuclear axis and therefore allow the transition between two different levels of vibration in the same quantum state of rotation, i.e. ΔJ=0. These are therefore called Q branches, an example of which is given on Figure 3.6, for the molecule HCN at approximately 710 cm^{-1}.

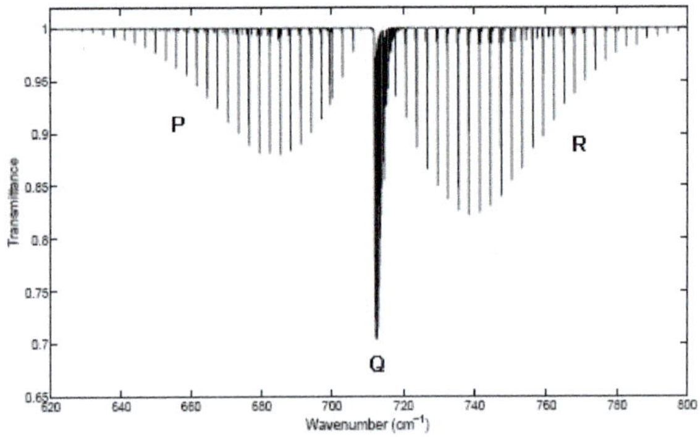

Figure 3.6. *Transmittance spectrum of the fundamental vibration mode v_2 of HCN. The branches P and R are located to the left and right, respectively. The lines most intense at the center belong to branch Q*

3.1.3. *Line shapes*

According to equation [3.1], if we disregard instrumentation influences, an observed transition should be mathematically represented by a Dirac peak. In reality, this is not the case and several effects contribute to the shape and spectral width of transitions. Note that if the experimental observation of a transition between two energy levels corresponds to a curve, we use the term "line". This term occurred during the first observations carried out on solar or flame spectra *via* gratings, which form vertical lines at different wavelengths.

3.1.3.1. *Natural broadening*

The relation that links the energy of a state to its radiative lifetime Δt is governed by the Heisenberg inequality:

$$\Delta E.\Delta t \geq \hbar, \qquad [3.30]$$

where $\hbar$ is the reduced Planck constant (or Dirac constant).

This equation illustrates the fact that a state can only have a perfectly defined energy if its lifetime is infinite. As this is not the case, there is a certain extension of energy levels and therefore their transitions, which will enlarge the lines observed.

We can associate this broadening $\Delta\nu$ to the moment of transition:

$$\Delta\nu = \frac{\Delta E}{h} = \frac{1}{2\pi\tau} \approx \frac{32\pi^3\nu^3}{(4\pi\varepsilon_0)3hc^3}|\mathcal{M}_{kl}|^2 \qquad [3.31]$$

The dependence of $\Delta\nu$ on ν^3 induces a significant broadening for an electronic state ($\Delta\nu \approx 30\text{ MHz}$) than for a rotational state ($\Delta\nu \approx 10^{-4} - 10^{-6}\text{ Hz}$). This value is therefore very low in the infrared (typically 10^{-10} cm^{-1}), and can be considered as negligible compared with other sources of broadening.

3.1.3.2. *Pressure broadening*

Collisions due to pressure between atoms or molecules cause distortions, with energy exchange and therefore a spread of energy levels. If an atom, which is being emitted/absorbed, is subject to a collision, the emission/absorption stops. If τ is the mean time between two successive collisions, the broadening is:

$$\Delta \nu = \frac{1}{2\pi\tau} \qquad [3.32]$$

The mean time τ can also be expressed using the kinetic theory of gases. If n is the number of particles per unit volume and σ is the cross section of collision, the mean free path between two collisions is equal to:

$$\lambda = \frac{1}{\sqrt{2}}\frac{1}{n\sigma} = \frac{1}{\sqrt{2}}\frac{kT}{\sigma}\frac{1}{p} \qquad [3.33]$$

The average time between two collisions is therefore:

$$\tau = \frac{\lambda}{\overline{V}} = \frac{1}{\sqrt{2}}\frac{kT}{\overline{V}\,\sigma}\frac{1}{p} \qquad [3.34]$$

where $\overline{V}$ is the average speed of molecules. The line-width is therefore proportional to pressure p. We can easily understand that the greater the pressure, the more collisions there are, and the line-width increases as less time is spent on one level.

With the distribution of lifetime between two collisions following the Poisson distribution, Lorentz showed that the line-shape could be described by equation:

$$f_L(\tilde{\nu}) = \frac{\alpha_L/\pi}{(\tilde{\nu}-\tilde{\nu}_0)^2+\alpha_L{}^2} \qquad [3.35]$$

where α_L is the half width at half maximum (HWHM), such that: $\alpha_L = \alpha_0(p/p_0)(T_0/T)^{1/2}$, α_0 being the width at half maximum for standard temperature and pressure, respectively p_0 and T_0.

Broadening by pressure (low troposphere <20 km) is typically around 10^{-3} cm^{-1}, i.e. several orders of magnitude greater than lifetime broadening.

3.1.3.3. *Doppler broadening*

During the process of absorption/emission, the frequency at which the transition occurs depends the speed the molecule is moving with respect to the detector. This effect is known as the Doppler effect. In the case where gases are very dilute (as in the upper atmosphere >50 km) collisions are rare, but molecules maintain a velocity associated with thermal agitation. Thus, the wavenumber at which the transition will be observed is as follows:

$$\tilde{\nu} = \tilde{\nu}_0 \,(1+v/c) \qquad [3.36]$$

with $\tilde{\nu}_0$ the central wavenumber of transition, c the speed of light and v the wave speed in the axis of the observer.

For a gas at thermodynamic equilibrium, the kinetic theory of gases describes the distribution of the speed of molecules using the Maxwell distribution resulting in the following Doppler profile:

$$f_D(\nu) = \frac{1}{\alpha_D\sqrt{\pi}} \exp\left(-\frac{(\tilde{\nu}-\tilde{\nu}_0)^2}{\alpha_D^2}\right) \qquad [3.37]$$

with α_D, half width at half maximum, such that:

$$\alpha_D = \frac{\tilde{\nu}_0}{C}\sqrt{\frac{2kT\, ln2}{m}} \qquad [3.38]$$

m being the molecular mass.

Compared to the Lorentz profile, the Doppler profile can be of the same order of magnitude depending on the conditions of p and T, but with a different spectral distribution, since it is greater at the center of the line and lower in the line wings (Figure 3.7) and is similar to a Gaussian function.

3.1.3.4. *Voigt profile*

When the gaseous phase is in an environment where the pressure and temperature both play a role, typically for the atmosphere between 20 and 50 km altitude, radiation is subject both to the effects of pressure and speed. From a spectral point of view, it becomes therefore necessary to take these effects into account in the line shape. The convolution of the Doppler and Lorentz profile is known as the Voigt profile and is written:

$$f_V(\tilde{v}) = \frac{1}{\pi^{3/2}} \frac{\alpha_L}{\alpha_D} \int_{-\infty}^{\infty} \frac{1}{(\tilde{v}'-\tilde{v}_0)^2+\alpha_L{}^2} \exp\left[-\frac{(\tilde{v}'-\tilde{v}_0)^2}{\alpha_D{}^2}\right] d\tilde{v}' \qquad [3.39]$$

Note that the above expression is not a simple analytical form and must be numerically approximated. This problem is well known for the calculation of line-by-line radiative transfer mainly because of the high computing time that it requires.

With regard to sounding of the atmosphere at the nadir line-of-sight, which sounds the whole of the vertical atmospheric column, the Voigt profile is most commonly used.

However, note that in some limited cases, collisions and the molecular movements cannot be considered as independent phenomena because they are linked via two different processes: (1) change in speed (Dicke effect) by collision, which results in a Galatry profile; (2) the dependence of relaxation rate with molecular speed, which results in a speed-dependent Voigt profile. Ultimately, both processes appear simultaneously; they are therefore modeled using the speed-dependent Galatry profile.

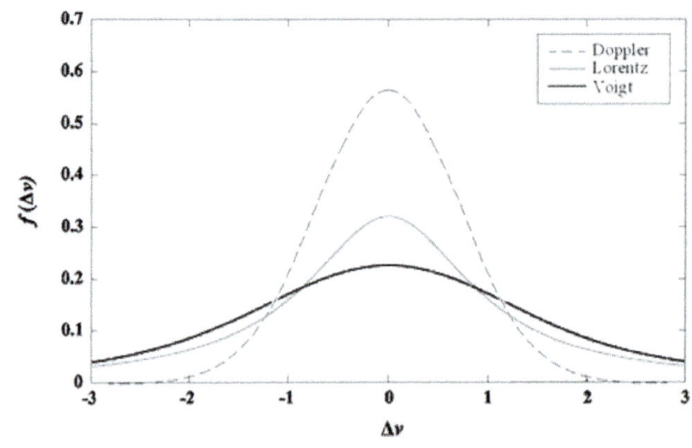

Figure 3.7. *Examples of line shape: the black curve corresponds to a Gaussian function (Doppler effect), the gray curve to a Lorentzian function (effect of pressure) and the dash curve represents the combination of both represented by a Voigt function*

3.1.4. *Line intensity and absorption coefficient*

The intensity [cm^{-1}/(molecule.cm^{-2})] of a transition between two vibration-rotation levels at the reference temperature T_{ref} = 296 K is written:

$$S_{kl} = \frac{h\nu_{kl} n_k}{CN}\left(1 - \frac{g_k n_l}{g_l n_k}\right)\mathcal{M}_{kl}{}^2 10^{-36}\frac{8\pi^3}{3h^2} \qquad [3.40]$$

where n_k and n_l are the populations of low and high states, respectively, g_k and g_l are the degenerate levels of k and l, respectively, and N is the molecular density.

The intensity distribution of transitions depends on the value of the square of the moment of transition ($\mathcal{M}_{kl}{}^2$) defined by equation [3.5], but also on the population of states involved. The latter is obtained from the Boltzmann distribution law. Assuming local thermodynamic equilibrium (LTE), the function of the population partition is governed by the Boltzmann statistical law:

$$\frac{N_k}{N_l} = \frac{g_k}{g_l}\exp\left(-\frac{\Delta E}{kT}\right). \qquad [3.41]$$

The intensities S_{kl} , at T_{ref} = 296 K and p_{ref} = 1 atm, are accurately determined in the laboratory and the intensity of transition at temperature T can then be calculated using the following relation:

$$S_{kl}(T) = S_{kl}(T_{ref})\frac{Q(T_{ref})}{Q(T)}\frac{\exp\left(-\frac{hCE_k}{kT}\right)}{\exp\left(-\frac{hCE_k}{kT_{ref}}\right)}\frac{\left[1-\exp\left(-\frac{hC\nu_{kl}}{kT}\right)\right]}{\left[1-\exp\left(-\frac{hC\nu_{kl}}{kT_{ref}}\right)\right]} \qquad [3.42]$$

where E_k is the energy of the lowest state [cm^{-1}] and $Q(T) = \sum_k g_k \exp\left(-\frac{hCE_k}{kT}\right)$ is the sum of the internal partition function.

The absorption coefficient [1/(molecule.cm^{-2})] of a given transition j is therefore the product of the previous intensity with the line shape (see section 3.1.3), such that:

$$k_j^{abs.}(\nu,T,p) = S_j(\nu,T,p).f(\nu,\nu_j,T,p) \qquad [3.43]$$

If several lines are spectrally very close, the total absorption coefficient k_v at v is therefore the sum of the absorption coefficients of each transition j multiplied by the quantity of absorbent u of each molecular species i considered, such that:

$$k_v = \sum_{i=1}^{u_g} u_i \sum_j k_j^{abs.}(v, T, p) \quad [3.44]$$

3.2. Radiative transfer equation in an absorbing medium

Remote sensing measurements obtained from radiometers or infrared spectrometers correspond to the intensity of electromagnetic radiation that has passed through the atmosphere. It is therefore not a direct measurement of the concentration of gases or particles in the atmosphere, but their effect on radiation. Modeling radiative transfer between the radiation source and the instrument is therefore essential in order to determine the spectral radiance observed according to the environmental properties.

Solving the radiative transfer equation (RTE) in a realistic atmosphere, that is to say, taking into account all the radiative processes and their interactions, is a complex problem because the distribution of atmospheric constituents is, in actual fact, not really homogeneous. Therefore, solving this requires the use of three-dimensional methods. However, a one-dimensional approach (vertical) is frequently used, for applications that require many simulations (climate modeling, operational processing of satellite data, etc.), because it is simple, and involves reasonable calculation times and gives satisfactory results in most cases.

Thus, we will consider the atmospheric environment to be stratified into horizontally infinite and parallel homogenous layers (see Chapter 4). Each layer is usually defined by the temperature and pressure limits and optical parameters. In this case, each atmospheric column is considered to be independent and the interactions of a column with another are not taken into account.

The formulation of the RTE in a purely absorbing medium is obtained from the radiative energy conservation balance of an incident wave through a volume element taking into account the processes of absorption and emission. In general, it derives from the Beer–Lambert relationship used to describe the evolution of the intensity of an electromagnetic wave when interacting with atmospheric layer along a path length *dz*, such that:

$$dI(\nu) = -k_{\nu}^{abs.} I(\nu)dz \qquad [3.45]$$

The atmospheric layer crossed also emits radiation, which can be written in the form:

$$dI(\nu) = k_{\nu}^{emis.} J(\nu)dz \qquad [3.46]$$

where $k_{\nu}^{emis.}(\nu)$ is the emission coefficient of the environment and $J(\nu)$ is a "source function".

In the atmosphere, up to altitudes of approximately 60 to 70 km, the density of molecules is such that the "thermal" emission (due to collisions between molecules) allows us to disregard the spontaneous emission of molecules. This therefore falls within the context of local thermodynamic equilibrium. In other words, only the temperature governs the distribution of molecules in different energy levels (equation [3.41]). Consequently, the radiation emitted by each layer only depends on its temperature T and $J(\nu)$ is given by Planck's function:

$$J(\nu) = B(\nu,T) = 2h\nu^3c^2[\exp(h\nu/kT) - 1] \qquad [3.47]$$

This function corresponds to the radiation emitted by a blackbody at a frequency ν and at temperature T. By applying the Kirchoff law, locally valid at each layer of the atmosphere, the coefficients of absorption $k_{\nu}^{abs.}$ and emission $k_{\nu}^{emis.}$ are equal and we obtain the following form for the radiative transfer equation:

$$dI(\nu)/dz = -k_{\nu}[I(\nu) - B(\nu)] \qquad [3.48]$$

Another form of this equation can be established by introducing the integrated gaseous transmission $T_{\Delta\nu}$ on a spectral interval $\Delta\nu$ and on a geometric path between two altitudes z_1 and z_2, which, for a given gas, is expressed as:

$$T_{\Delta v} = \frac{1}{\Delta v} \int_{\Delta v} dv \int_{z_1}^{z_2} \exp[-k_v(z)]\, dz, \qquad [3.49]$$

By integrating the length of the optical path, we obtain the general formulation for the RTE:

$$I(v,z) = I_0 T_{\Delta v} + \int_{z_1}^{z_2} k_v(z) B(v,T)\, exp[-k_v(z)]\, dz \qquad [3.50]$$

where I_0 corresponds to the radiation emitted by the light source in $z = 0$ (blackbody on the surface of the Earth or Sun). Therefore, from the previous equation, the different forms for the spectral radiance can be obtained, in the case of a purely absorbing atmosphere according to the line of sight, such that:

– *Limb viewing*:

$$L_{Limb} = \int_{Lfar}^{Lnear} k_v(s) B(v,T). \exp\left(-\int_S^{Lnear} k_v(s) dS'\right) dS \qquad [3.51]$$

where S represents the optical path between the nearest point in the atmosphere L_{near} of the instrument and the farthest L_{far}.

– *Solar occultation*:

$$L_{SO} = \int_{Lfar}^{Lnear} k_v(s). B(v,T). \exp\left(-\int_S^{Lnear} k_v(s) dS'\right)\ dS + L_{solar} \exp\left(-\int_{Lfar}^{Lnear} k_v(s) dS'\right) \qquad [3.52]$$

with L_{solar} the solar spectrum at the top of the atmosphere.

– *Zenith viewing (or ground measurement)*:

$$L_Z = \int_0^H k_v(h) B(v,T) \exp\left(-\int_0^h k_v(h) dh'\right) dh + I_{solar} \exp\left(-\int_0^H k_v(h) dh\right) \qquad [3.53]$$

where h represents the altitude of atmospheric levels between the surface (z_1=0) and the top of the atmosphere (z_2=H).

– *Nadir viewing*:

$$L_N = \varepsilon B(T_0)\exp\left(-\int_0^H k_v(h)dh\right) + (1-\varepsilon)L_Z\exp\left(-\int_0^H k_v(h)dh\right) + \int_0^H k_v(h)B(v,T)\exp\left(-\int_h^H k_v(h)dh'\right)dh \quad [3.54]$$

where ε and T_0 are the emissivity and the surface temperature, respectively.

The first term of equation [3.54] corresponds to the radiation emitted by the Earth's surface (part 1 of Figure 3.8) transmitted through all atmospheric layers to the satellite. The second term represents the part emitted by the atmosphere and the solar radiation transmitted that is reflected by the surface (parts 2a and 2b of Figure 3.8). The third is associated with the radiation emitted by the atmospheric layers and transmitted in the direction of the satellite (part 3 of Figure 3.8).

Note that the formulation of the previous equations does not include the directional aspect, which is essential in the case of a significant angle of sight (dh is therefore replaced by $sec(\theta)$ dh, with θ the angle of the direction of observation with regard to the vertical).

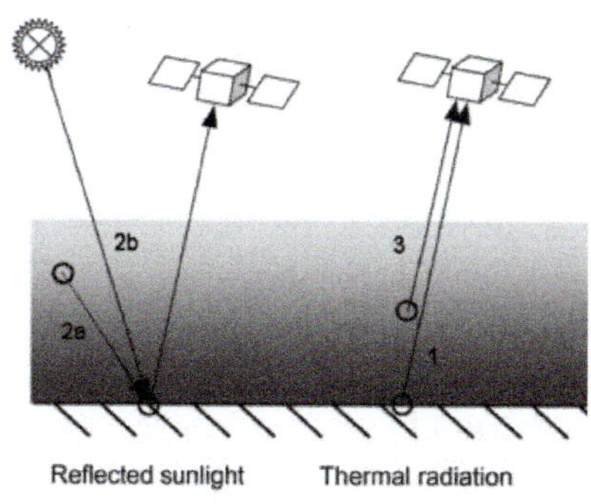

Figure 3.8. *Illustration showing the observation of the radiation coming from the Earth's surface (1), the Sun (2b) and the atmosphere (2a and 3) from the Nadir viewing*

3.3. Solving the RTE

Two approaches are used to calculate the radiance according to the desired accuracy: (1) the high spectral resolution calculation called "Line-By-Line" (LBL) is the most accurate solution, but requires significant computation time when there is a wide spectral range; (2) approximate models with coarser spectral resolution, known as "band models", are alternative solutions for faster calculations or when high resolution is not required.

3.3.1. *Models at high spectral resolution: line-by-line codes*

For a given interval $\Delta\nu$, the LBL approach involves calculating the attenuation coefficient $k_j^{abs.}$ of each gaseous species at all levels of the atmosphere, with a very narrow spectral resolution, less than 0.01 cm^{-1}. Then to simulate the radiance as measured by the instrument L^{conv}, we must convolve $L(\nu)$ calculated at high spectral resolution using the instrumental function (ILS) g, such that:

$$L^{conv}(z) = \int_{v-\Delta v}^{v+\Delta v} L(v')g(v'-v)dv' \quad [3.55]$$

The spectroscopic parameters required for a LBL calculation, such as the spectral position of lines, their intensity or their width at half-maximum are obtained from laboratory measurements and/or theoretical calculations from quantum mechanics and statistical physics. All data is available in databases, the most well-known being GEISA and HITRAN:

– *GEISA* (Gestion et Etude des Informations Spectroscopiques Atmosphériques) as developed by the group ARA (Atmospheric Radiation Analysis) at the laboratory LMD (Laboratoire de Météorologie Dynamique) in France. It was disseminated by the Pôle de Compétence Thématique Ether: http://ether.ipsl.jussieu.fr/etherTypo/?id=950 [JAC 11].

– *HITRAN* (high-resolution transmission molecular absorption database) is the result of research carried out since the 1960s by Air Force Cambridge Research Laboratories (AFCRL), and continued today by the Atomic and Molecular Physics Division, Harvard-

Smithsonian Center for Astrophysics (USA): http://www.cfa.harvard.edu/hitran/ [ROT 13].

Another challenge with the gas absorption calculation comes from considering the distant line wings which, through the effect of accumulation, contribute to a significant absorption in some spectral ranges for which the number and the intensity of lines are significant. This is particularly true for water vapor in the thermal infrared range of the spectrum, with the accumulation distant line wings generating a continuous absorption spectrum called an "absorption continuum." Figure 3.9 highlights the contribution in the infrared atmospheric window. Accurately taking into account these continua present two major problems: (1) first of all, theoretical, since very little is known about the exact shape of the distant line wings; (2) then practical, since taking into account these line wings requires significant calculation times. In practice, their impact is taken into account in semi-empirical parameterization, based both on theory and on spectroscopic measurements. For the water vapor continuum, the continuum coefficients were calculated across the infrared spectrum by Clough *et al.* [CLO 89]: CKD parameterization (http://rtweb.aer.com/continuum_description.html). Note that other continua also have a significant contribution in the infrared (CO_2, N_2, O_3, etc.) and the corresponding absorption coefficients are also available in the CKD parametrization.

3.3.2. *Approximate modeling of gas absorption*

3.3.2.1. *Band models*

The first models of gaseous absorption were based on the consideration of isolated absorption lines such as the regular model of Elsasser [ELS 60], using simple analytical relations involving the average line width for Lorentz or Doppler line shapes. However, these regular band models quickly revealed their limitations, especially their difficulty when taking into account a complex Voigt line shape and line overlap.

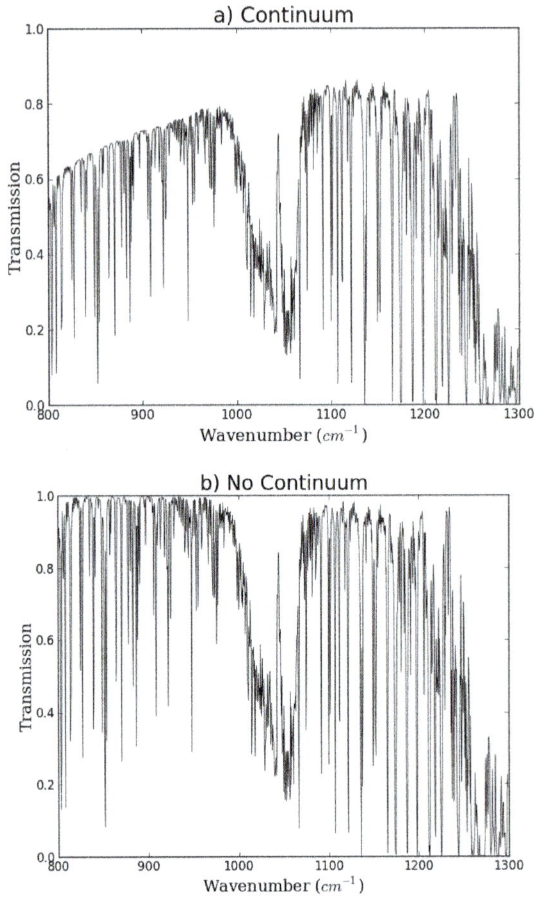

Figure 3.9. *Atmospheric spectral transmission calculated in the infrared window using an LBL code for a standard Mid Latitude Summer atmosphere by considering a) the water vapor continuum or b) when this continuum is not taken into account*

A more accurate approach is to use random band models. In fact, although the absorption lines of gaseous species are organized into absorption bands, the hypothesis of isolated and evenly spaced lines is unrealistic and leads in most cases to significant errors. Random band models have been developed to take into account a probabilistic distribution of lines in a spectral range $\Delta\nu$. The mean transmission $T_{\Delta\nu}$

for N randomly spaced lines by an average distance δ such that $\Delta\nu = N\delta$ is then expressed as follows:

$$T_{\Delta\nu} = \prod_{i=1}^{N} \frac{1}{\Delta\nu} \int_{\Delta\nu}\int_{0}^{\infty} p(S_i) \exp\left[-uS_i g(\nu - \nu_i)\right] dS_i d\nu_i . \quad [3.56]$$

The term $P(S_i)dS_i$ defines the probability that the intensity of a line i is between S_i and $S_i + dS_i$. Two models are generally used:

– The Goody model, developed by assuming a Poisson distribution for the distribution of line intensities S:

$$T_{\Delta\nu} = \exp\left[-u\frac{\overline{S}}{\delta}\left(1 + \frac{u\overline{S}}{\pi\overline{\alpha}}\right)^{-1/2}\right]. \quad [3.57]$$

– The Malkmus model, which uses an S^{-1} distribution, giving a stronger probability for weak lines:

$$T_{\Delta\nu} = \exp\left[-\frac{2\pi\overline{\alpha}}{\delta}\left\{\left(1 + \frac{u\overline{S}}{\pi\overline{\alpha}}\right)^{1/2} - 1\right\}\right]. \quad [3.58]$$

For these two expressions, the mean values of intensity and width at half-maximum of lines are respectively defined as follows:

$$\overline{S} = \frac{1}{N}\sum_{i=1}^{N} S_i \quad ; \quad \overline{\alpha} = \frac{4}{\pi\overline{S}}\left[\frac{1}{N}\sum_{i=1}^{N}\left(S_i\alpha_i\right)^{1/2}\right]^2. \quad [3.59]$$

These transmissions can be easily calculated for a temperature T and pressure p, using a spectroscopic database such as GEISA or HITRAN, which provides access to line parameters S_i and α_i. However, they do not take into account the spectral shape or line overlap.

These models are well suited for calculating the absorption for a homogeneous path, at a fixed pressure and temperature. In the case of a vertically inhomogeneous atmosphere, using scaling approximations with one or two parameters helps account for variations in pressure and temperature. These methods include defining a homogeneous environment as equivalent to real one, with equivalent values for pressure p_e, temperature T_e and quantity of absorbent u_e. The aim is to fix one (scaling approximation) or two (Curtis–Godson) of these parameters, with the others being determined so as to retrieve the absorption with acceptable accuracy, for both strong and weak absorption regimes [THO 02]. These approximations are still useful due to their simplicity and educational value since they allow the absorption to be calculated using simple analytical expressions.

3.3.2.2. *k-distribution method*

This method is fast and accurate and is an interesting alternative to the LBL approach. Originally based on the work of Ambartzumian [AMB 36] and Lebedinsky [LEB 39], nowadays it is widely used for calculating radiative transfer and a detailed description is provided by [LAC 91]. The main idea lies in the fact that, for a given Δν, the same absorption coefficient may occur several times and that the exact spectral position is not important when calculating the mean transmission $T_{\Delta\nu}$. For a given quantity of absorbent u, a temperature T and pressure p, $T_{\Delta\nu}$ is defined using the fraction $f(k)$ of Δν corresponding to an absorption coefficient between k and $k+dk$ such as:

$$T_{\Delta\nu}(p,T)=\frac{1}{\Delta\nu}\int_0^\infty f(k)\exp[-k(p,T)u(p,T)]dk, \text{ with} \int_0^\infty f(k)dk=1\cdot$$

From $f(k)$, we define the cumulative frequency distribution g such that $g(k_l)=\int_0^{k_l} f(k)dk$, with $g(0) = 0$ and $g(\infty) = 1$. Figure 3.10(b) gives an example of a cumulative frequency distribution g of line absorption coefficients of water vapor (presented in Figure 3.10 (a)) for the spectral region between 3600 and 4000 cm^{-1}.

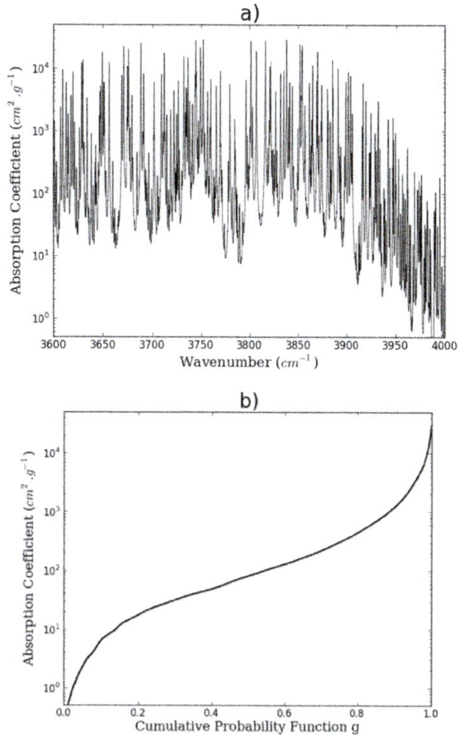

Figure 3.10. *a) Spectral variations in the absorption coefficient k and b) cumulative frequency distribution g of the absorption coefficient, under standard conditions of pressure and temperature*

The regular nature of function g allows the discretization of the absorption coefficient space into a number N of absorption classes k_i, each associated with a weight in the spectral interval. The mean transmission Δν can therefore be approximated using a finite sum such that:

$$T_{\Delta\nu}(p,T) \approx \sum_{i=1}^{N} \Delta g_i \exp[-k_i(p,T)u(p,T)] \approx \sum_{i=1}^{N} a_i \exp[-k_i(p,T)u(p,T)] \qquad [3.60]$$

The weights a_i of the distribution represent the probability associated with the absorption coefficient k_i such as:

$a_i = \int_{k_{l-1}}^{k_l} f(k)dk = g(k_l) - g(k_{l-1})$, with $\sum_{i=1}^{N} a_i = 1$. From this formulation, the RTE is solved for each absorption class, the results being weighted by the coefficients a_i. For example, the total radiative flux $F_{\Delta\nu}$ is calculated over Δν as the sum of flux $F(k_i)$ obtained for each absorption class: $F_{\Delta\nu} = \sum_{i=1}^{N} a_i F(k_i)$.

For a given temperature T and pressure p, the coefficients $[a_i, k_i]$ can be easily calculated using a LBL code, for a reduced number of coefficients N, thereby allowing rapid calculations of the transmission over the given spectral interval. It also takes into account the spectral response of an instrument $\Phi(\nu)$, whose integrated response is $R_\Phi = \int_{\Delta\nu} \Phi(\nu)d\nu$ [WEN 12]. Now let us assume that the spectral interval Δν is divided into $m = 1,M$ spectral sub-intervals, each with a width of $\delta\nu_m$ and corresponding to a value Φ_m for the spectral response and k_m for the absorption coefficient. Furthermore, the maximum and minimum values of $k(\nu)$ are calculated from the LBL model over Δν, defining the complete range of absorption coefficients, which is divided into $i=1,N$ sub-intervals δk_i such that $k_i = k_{i-1} + \delta k_i$. The cumulative frequency distribution g is therefore calculated by classing each absorption coefficient k_m into an appropriate interval for which $k_{i-1} < k_m \leq k_i$ such that:

$$g(k_i) = \frac{1}{R_\Phi} \sum_{m=1}^{M} \Phi_m \delta\nu_m W_m(k_m \leq k_i), \qquad [3.61]$$

with W_m a function associated to the sub-interval m equal to 1 if $k_m \leq k_i$ and 0 if not.

For a vertically inhomogeneous atmosphere, the most common approach is to calculate the absorption coefficients assuming they are correlated with altitude, which means that the maximum absorption in a spectral range will always occur at the same wavenumber, whatever the p and T. In this case, the coefficients a_i are constant with regard to

p and T for a given spectral interval. This hypothesis, known as the "correlated k-distribution", generally gives good results and is widely used. For a given spectral interval, the coefficients k_i are pre-calculated for a reference pressure and temperature and they are then interpolated based on the actual conditions of each atmospheric layer. The mean transmission between two altitudes z_1 and z_2 is therefore expressed as:

$$T_{\Delta\nu}\left(z_1,z_2\right)=\sum_{i=1}^{N}a_i\prod_{z=z_1}^{z_2}\exp\left[-k_i(z)\,u(z)\right]. \qquad [3.62]$$

This approach helps avoid errors caused by the non-multiplicity of absorption encountered with band models such as Goody or Malkmus models. Note that more complex "non-correlated k-distribution" approaches, however, give more accurate results [DOP 14]. If several absorbing gases occur in the same spectral range, the mean transmission is usually calculated assuming that the two gases are not correlated. For two gases A and B, this calculation gives:

$$T_{\Delta\nu}(A,B)=T_{\Delta\nu}(A)\times T_{\Delta\nu}(B)=\sum_{i=1}^{N}\Delta g_i^A\sum_{j=1}^{M}\Delta g_i^B\exp\left(-k_i^A u^A-k_i^B u^B\right), \qquad [3.63]$$

with u_A and u_B the quantity of gas absorbent. This calculation requires $M\times N$ operations to obtain the mean transmission over $\Delta\nu$.

3.3.3. *Boundary conditions and atmospheric parameters*

Solving the RTE involves defining the boundary conditions of the atmospheric system and the atmospheric profiles of pressure, temperature and concentration of gases:

– at the TOA, the incoming source of radiation is the sun, defined from the extraterrestrial solar spectral flux $F_{0\nu}$ [W.m^{-2}.cm^{-1}] (see Chapter 1, section 1.1.2). These spectral fluxes have firstly been calculated in the visible range of the spectrum by Neckel and Labs in 1984 with a resolution of approximately 100 cm^{-1} [NEC 84]. More recently, Kurucz [KUR 00] calculated the fluxes for the whole spectrum with a resolution of 1 cm^{-1} (http://kurucz.harvard.edu/sun.html). Finally, in 2003 Thuillier *et al.* proposed very high-resolution spectra. In the infrared range, however, the Planck emission from space is negligible;

– at the BOA, solar and terrestrial contributions of radiation are to be considered for land and sea surfaces. In the thermal infrared, they emit according to Planck's law with a temperature between 180 K and 330 K and as a function of surface emissivity ($\varepsilon_\nu = 1 - \varpi_{0\nu}$). In practice, the surface temperature may be very different to that of the atmosphere situated just above. If T_s is not known, it is common to make it equal to that of the atmosphere situate just above the surface. In the solar spectrum, the outgoing radiation reflected by the surface is defined using the directional reflectance of the surface ρ_g, defined by its BRDF, or by its albedo for a Lambertian surface (see Chapter 1, section 1.4.4);

– solving the RTE also requires that the profile of pressure and temperature at altitude is defined as well as the concentrations and optical properties of gases, aerosols and clouds at each layer of the atmosphere. These profiles allow the calculation of the mean values of these variables for each layer. Profiles of pressure, temperature and concentration of gases from the models or *in situ* measurements (atmospheric radio soundings) can be used.

Some typical values for surface properties, atmospheric profiles and atmospheric particles are given in Chapter 1. More precise data are also available and examples of databases are shows in Table A.4 of the Appendix.

3.4. For further information

[BER 05] BERNATH P.F., *Spectra of Atoms and Molecules*, OUP, 2005.

[CHA 60] CHANDRASEKHAR S., *Radiative Transfer*, Dover Publications Inc., New York, 1960.

[GOO 95] GOODY R.M., YUNG Y.L., *Atmospheric Radiation: Theoretical Basis*, 2nd ed., Oxford University Press, 1995.

[HER 66] HERZBERG G., *Electronic Spectra of Polyatomic Molecules,* Van Nostrand, New York, 1966.

[HER 89a] HERZBERG G., *Molecular Spectra and Molecular Structure: I. Spectra of Diatomic Molecules,* Krieger, 1989.

[HER 89b] HERZBERG G., *Molecular Spectra and Molecular Structure: II. Infrared and Raman Spectra of Polyatomic Molecules*, Krieger, 1989.

[HOL 04] HOLLAS J.M., *Modern Spectroscopy*, John Wiley & Sons, 2004.

[LEF 86] LEFEBVRE-BRION H., FIELD R.W., *Perturbations in the Spectra of Diatomic Molecules*, Academic Press Inc, 1986.

[LEN 93] LENOBLE J., *Atmospheric Radiative Transfer*, A. Deepak Publishing, Hampton, VA, 1993.

[LIO 02] LIOU K.-N., *An Introduction to Atmospheric Radiation*, 2nd ed., Academic Press, 2002.

[PET 06] PETTY G.W., *A First Course in Atmospheric Radiation*, Sundog Publishing, Madison, Wisconsin, 2006.

[WEN 12] WENDISCH M., YANG P., *Theory of Atmospheric Radiative Transfer*, 1st ed., Wiley-VCH, 2012.

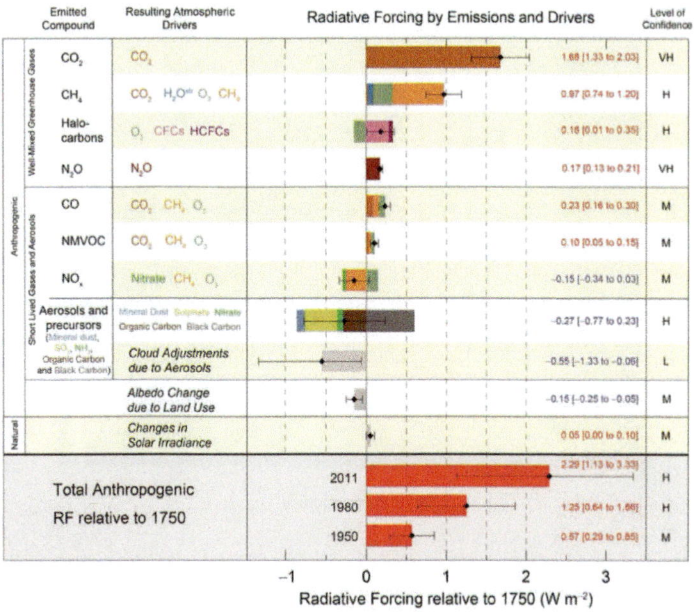

Figure 1.14. *Radiative forcing by concentration change between 1750 and 2011 with associated uncertainty range. Source: [IPC 13]*

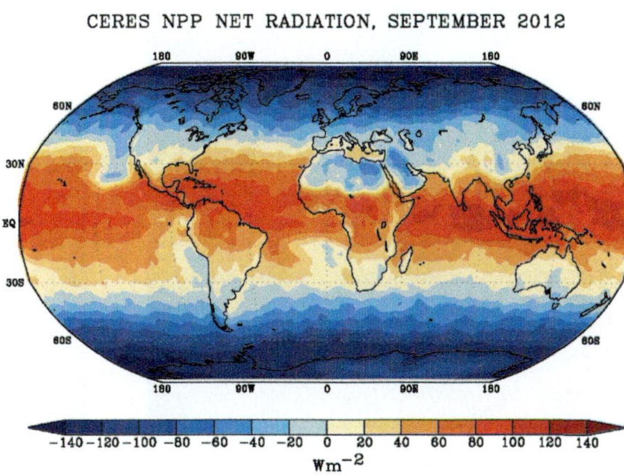

Figure 1.15. *Global Image of Net Radiation at Top-of-Atmosphere from CERES Instrument on NASA Suomi NPP Satellite for September 2012 (Instrument: CERES/NPP; Release Date: 12/7/2012). Source: NASA/CERES*

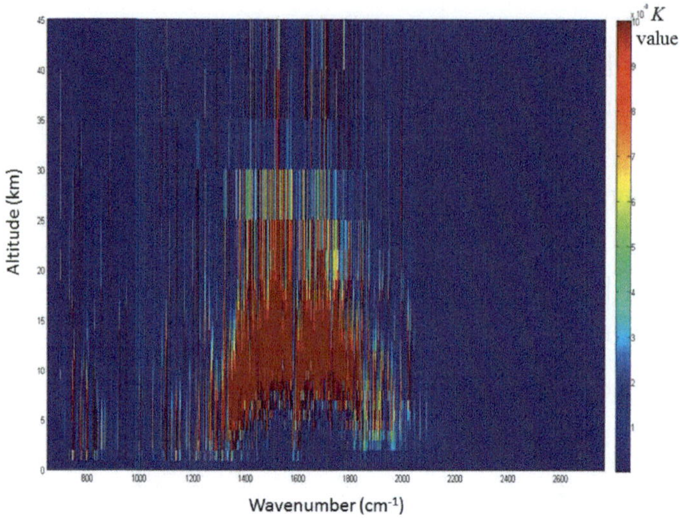

Figure 5.1. *Representation of the Jacobians K for $H_2{}^{16}O$ between 650 and 2760 cm^{-1} for a standard US profile, and the instrumental characteristics of the IASI sounder*

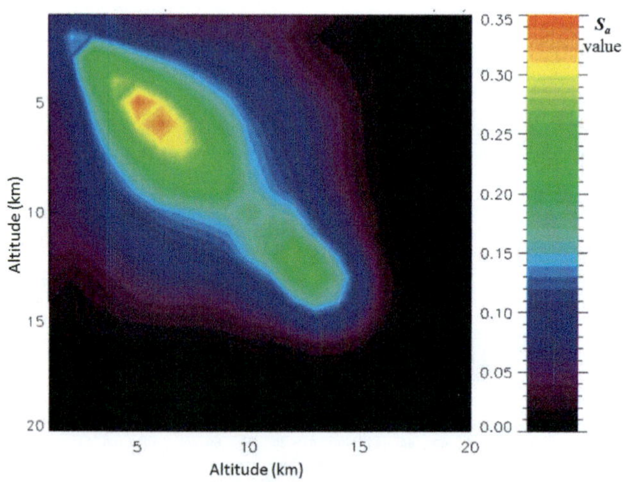

Figure 5.2. *Representation of the normalized covariance matrix S_a as a function of the altitude*

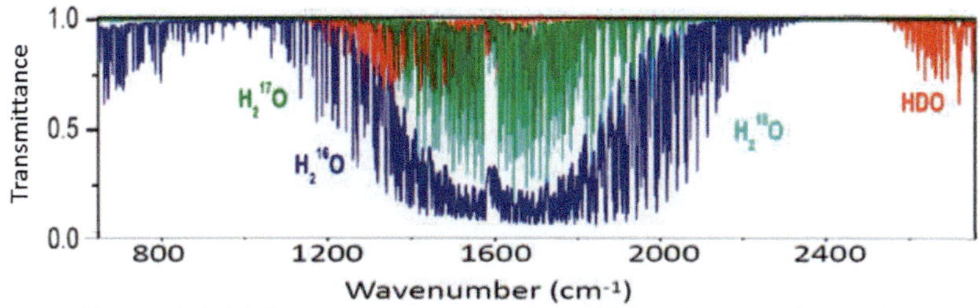

Figure 6.1. *IASI transmittance spectrum illustrating the absorption bands of the isotopologues: $H_2{}^{16}O$, $H_2{}^{18}O$, $H_2{}^{17}O$ and HDO (from [HER 07])*

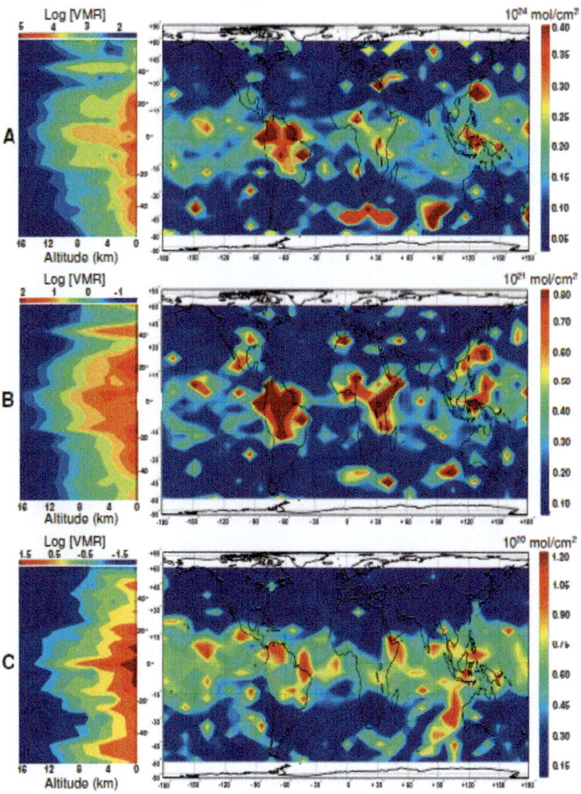

Figure 6.2. *Latitudinal distributions of the volume mixing ratio (VMR) [ppm] as a function of the altitude and latitude (left) and horizontal distributions on the global scale (right) of the concentrations [mol.cm^{-2}] of $H_2{}^{16}O$ A), $H_2{}^{86}O$ B) and HDO C) (from [HER 09])*

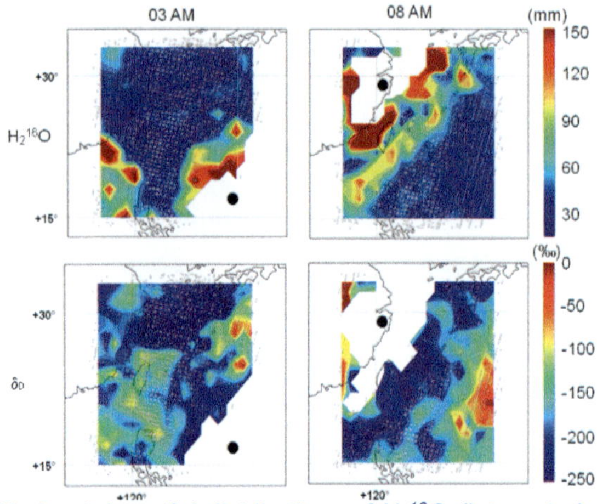

Figure 6.3. *Horizontal spatial distributions of $H_2{}^{16}O$ (integrated column along the vertical between 0 and 8 km of altitude in mm of precipitable water) and δD [‰] for the mornings of 3 and 8 October 2007. The black disk represents the eye of the typhoon. The white parts correspond to the cloudy areas. The gray dots represent the IASI IFOVs. The maps are averaged over a latitude/longitude grid of 1° (from [HER 09])*

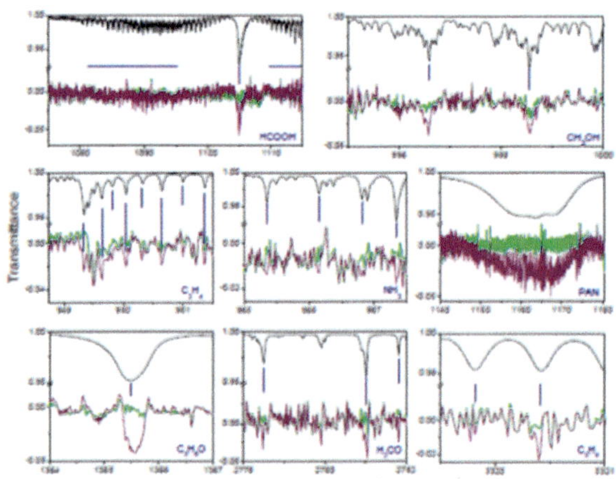

Figure 6.5. *Spectral adjustment of one ACE-fts spectrum corresponding to atangential altitude of 11 km, illustrating the detection of weakly absorbing compounds: CH_3OH, $HCOOH$, NH_3, C_2H_4, H_2CO, C_3H_6O, PAN and C_3H_4. The green and purple colors represent the residual, after adjustment, respectively with and without account taken of the absorption of the species in question. The spectral contribution specific to each species is shown in black, and the vertical blue lines mark the position of the main absorption lines*

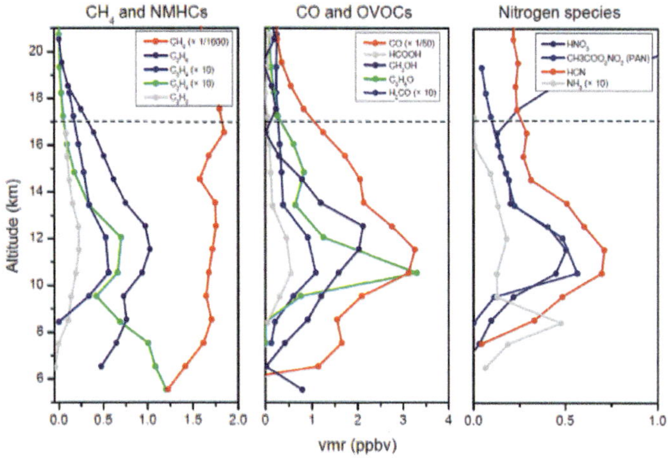

Figure 6.6. *VMR's vertical profiles (5–20 km) of the gaseous species given off by the biomass burning event. The horizontal dashed line marks the altitude of the tropopause*

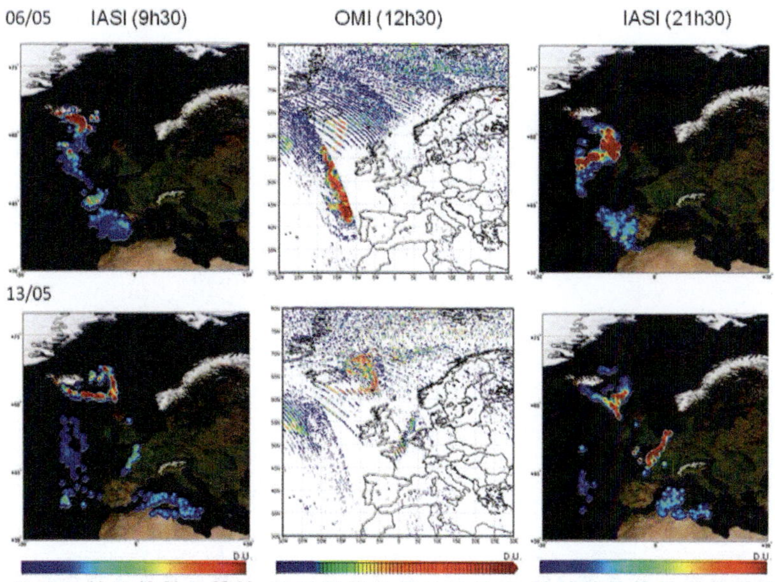

Figure 6.7. *Spatial distributions of the SO_2 total columns obtained by IASI (left and right) and OMI (middle) for 6 and 13 May. The IASI data are interpolated by a Cressman function (with R = 96 km). The OMI data are represented from the IFOV of the instrument*

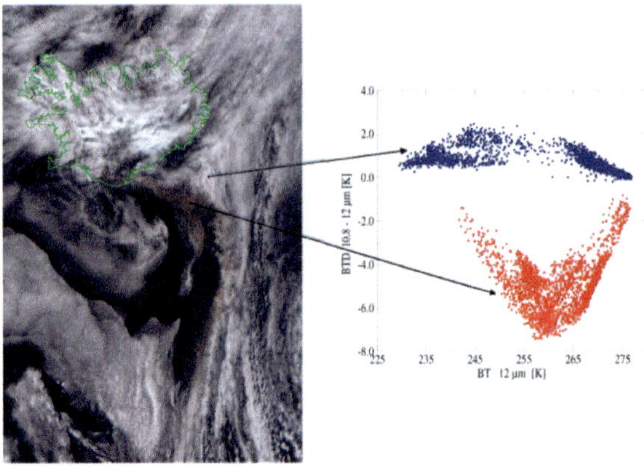

Figure 6.8. *Example of brightness temperature differences ΔBT [K] (right) during the eruption of the volcano Eyjafjallajökull, taken with the MODIS/Terra (left) on 6 May 2010. The representative values of the volcanic ash plume are shown in red; those in blue correspond to a cloudy area. ΔBT is defined as the difference between the brightness temperature at 10.8 and 12 micrometres. Source: [DUB 14]*

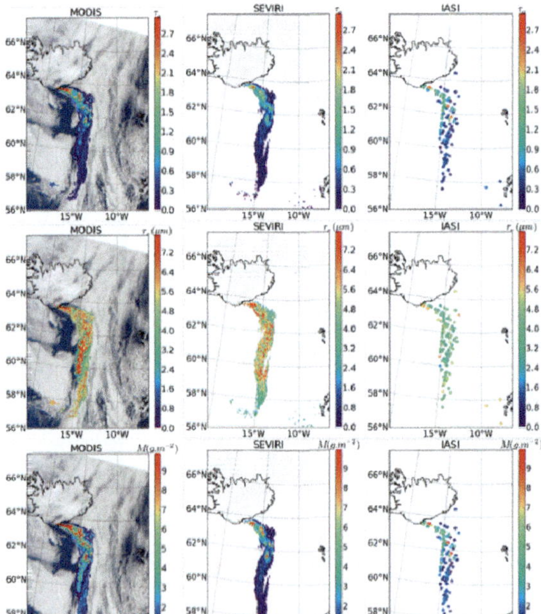

Figure 6.9. *Optical thickness τa at 12µm (top row), mean radius re (middle row) and surface mass M (bottom row) obtained on 6 May, using MODIS/Terra (11:55 UTC), SEVIRI (12:00 UTC) and IASI (late morning)*

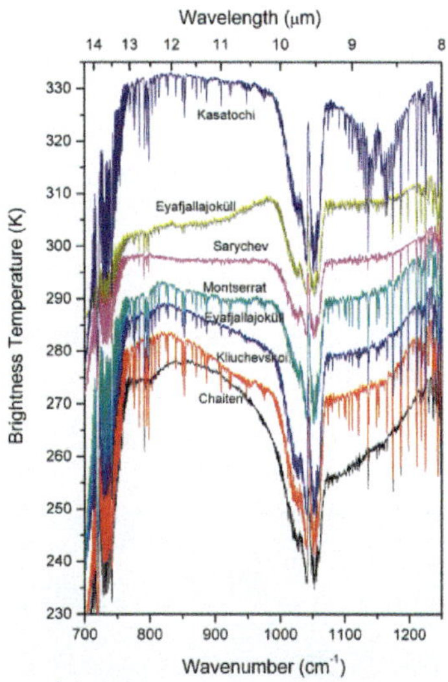

Figure 6.10. *Examples of brightness temperature spectra observed by IASI during several volcanic eruptions, illustrating the spectral variability of the plumes due to the diversity of the distributions of the particles emitted in terms of their size distribution (PSD) and type (chemical and mineralogical composition). These spectra are staggered on the ordinate (vertical) axis for ease of viewing. Adapted from [CLA 10]*

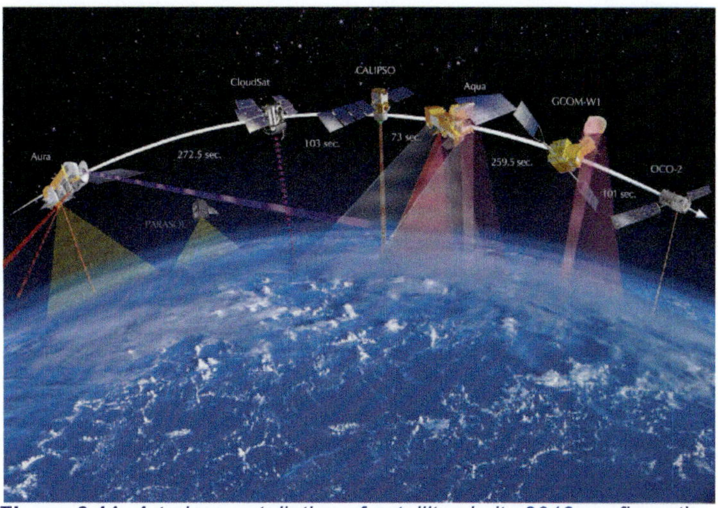

Figure 6.11. *A-train constellation of satellites in its 2013 configuration. Source: "A-Train w-Time2013 Web" by NASA JPL*

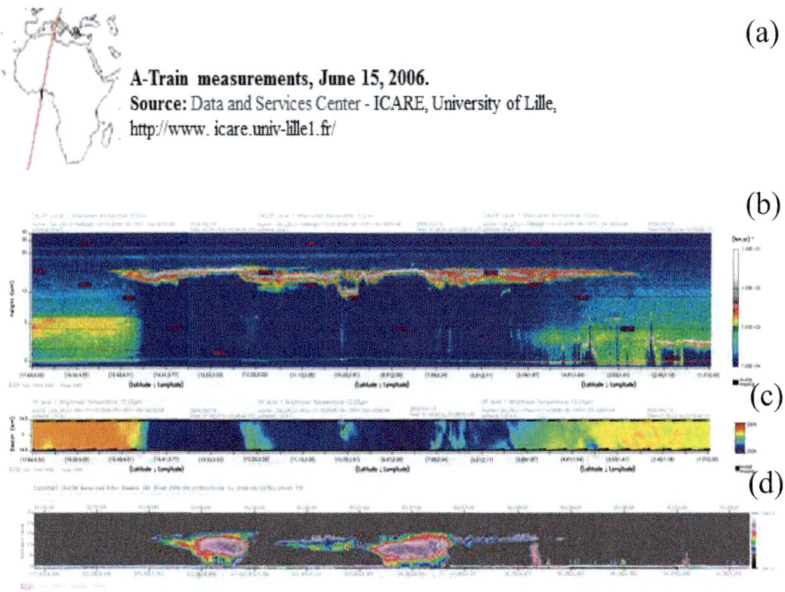

Figure 6.12. *Observation of a convective cloud system on 15 June 2006 above the Gulf of Guinea by the A-Train satellites: a) the red line represents its ground track, and the black line corresponds to the observed scene. This figure presents, as a function of the latitude/longitude; b) the vertical profile of the back scattering coefficient measured by CALIOP, showing high semi-transparent clouds around 15 km of altitude, and low aerosols and clouds around the edge of the cloud system; the red/gray zones indicate a high density of particles, the green/yellow zones a lower concentration and the dark blue zones indicate that the backscattered signal is null; c) the brightness temperature measured at 12 µm by IIR along its swath, cooler (blue) for high clouds and warmer (orange/yellow) for the aerosols and low clouds; d) the vertical profile of the measurements of CLOUDSAT/CPR, sensitive to the liquid-water and ice content (red/pink for high concentrations) of the cloud located between 1 and 12 km, below the semi-transparent cirrus clouds*

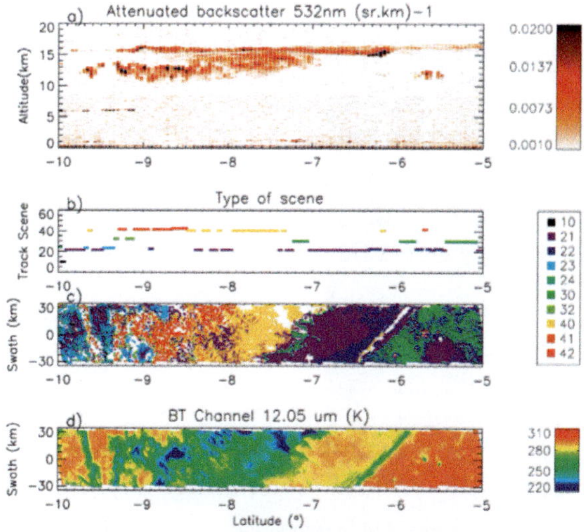

Figure 6.13. *Example of cloud classification as a function of the latitude for a scene observed by IIR: (a) vertical profile of the backscattering coefficient measured by the LIDAR CALIOP; (b) cloud classification corresponding to the above LIDAR profile, the color code for which is explained in Table 6.3; (c) classification extended to the whole of the plot observed by IIR; (d) brightness temperature at 12 µm along the length of the observed plot. Source: [GAR 12]*

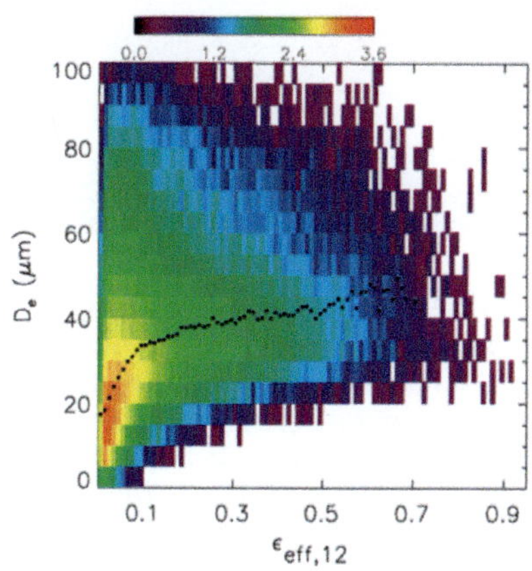

Figure 6.14. *Effective diameter D_e found retrieved using the IIR measurements as a function of the effective emissivity at 12 µm $\varepsilon_{eff,12}$ for single-layer cirrus clouds, between the latitudes 60°S and 60°N during the month of January 2011. The color code represents the point density and the black dotted line shows the median value for D_e, which lies approximately between 20 and 40 µm depending on the value of $\varepsilon_{eff,12}$. Source: [GAR 13]*

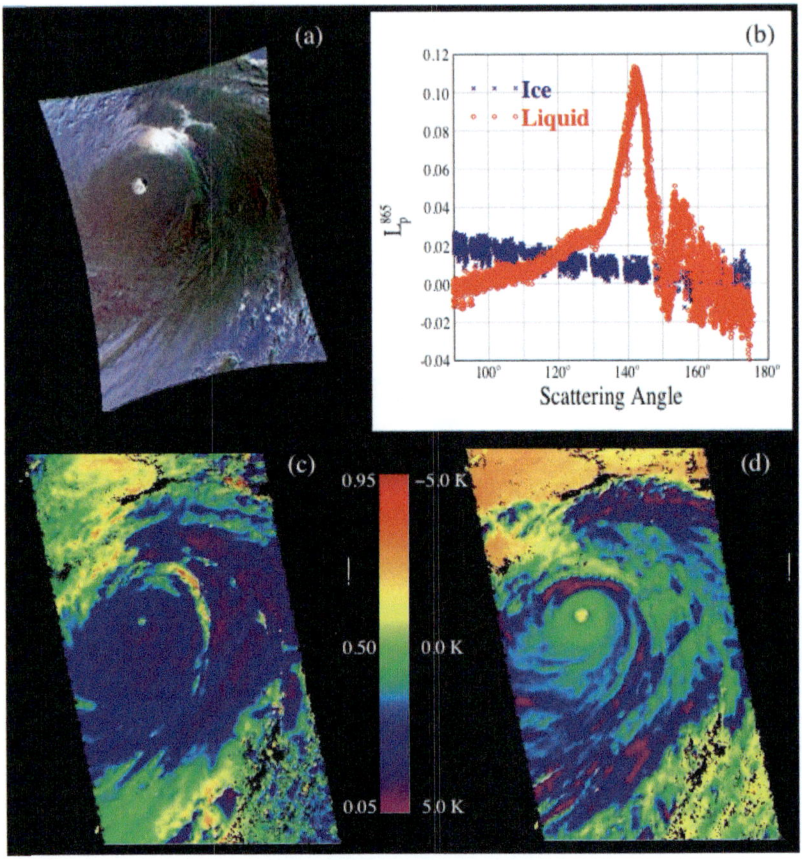

Figure 6.15. *Illustration of methods using MODIS and POLDER data to determination of the thermodynamic phase of clouds for a scene during Typhoon Nabi, on 2nd September 2005 off the coast of Japan: (a) scene used, artificially colorized using the polarized POLDER measurements at 490, 670 and 865 nm; (b) typical signature of polarized multi-angular radiance measured by POLDER for a cloud of liquid water (blue) or ice (red) as a function of the scattering angle; (c) ratio of MODIS measurements in the near-wave infrared (2.1 µm) and visible (0.865 µm); (d) brightness temperature difference (BTD) of the MODIS channels at 8.5 and 11 µm. Source: [RIE 10]*

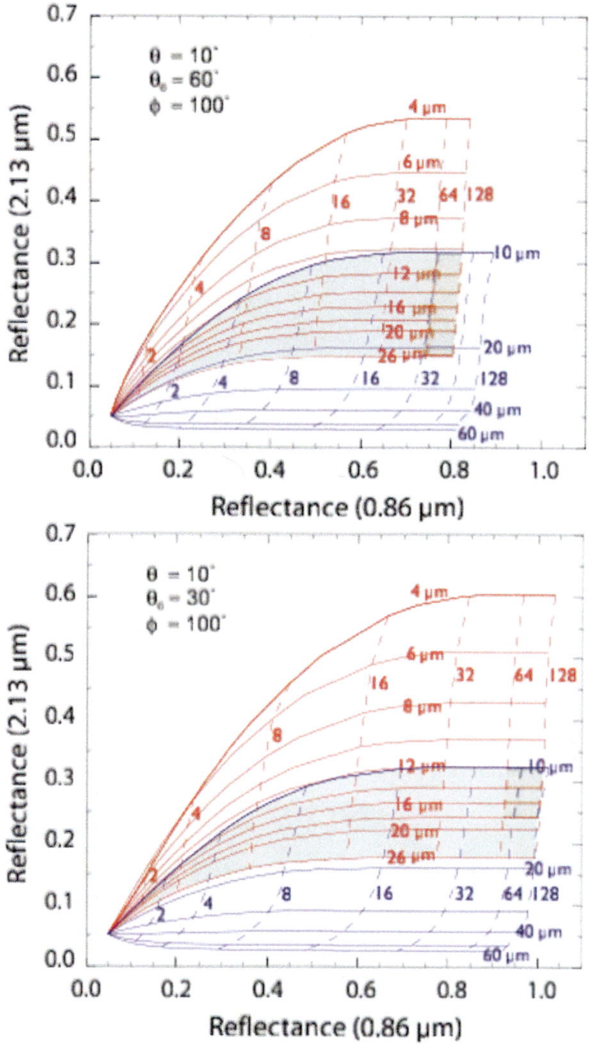

Figure 6.16. *Simulation of the reflectances at 2.13 µm as a function of those at 0.86 µm for two geometries of observation (solar zenith angle θ_0, observation angle θ and azimuth angle ϕ), for an ice cloud (blue) or liquid cloud (red), for different optical thicknesses (ranging between 0 and 128) and effective diameters of the particles (4-26 µm). The shaded area in the figure indicates that the combination of those wavelengths does not enable us to tell the phases apart. Source: [RIE 10]*

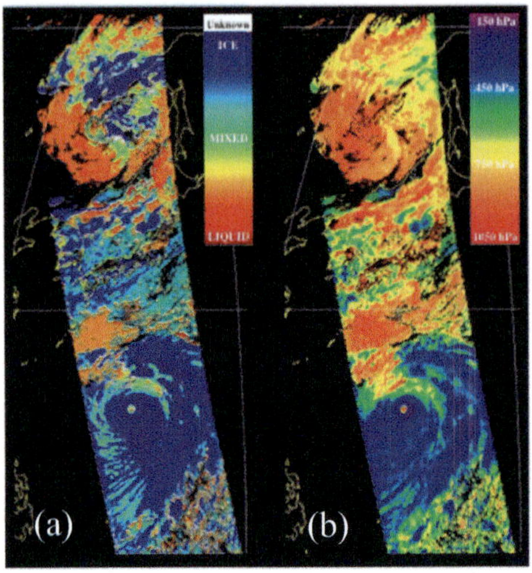

Figure 6.18. *(a)Classification of the cloudy phase by combination of measurements from POLDER and MODIS; (b) determination of the cloud top pressure using measurements in the POLDER channels centered in the dioxygen band around 760 nm, for the cloudy scene on 2 September 2005 presented in Figure 6.15(a). Source: [RIE 10]*

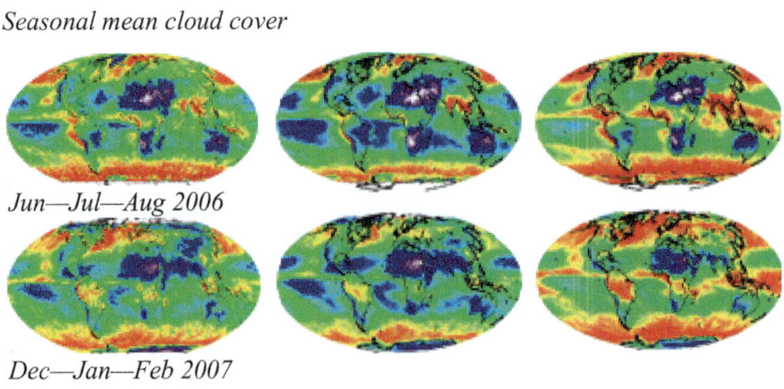

Figure 6.19. *Degree of cloud cover (from 0 to 1) deduced from measurements taken from POLDER (left-hand column), from MODIS (right-hand column) or the combination of the measurements taken by POLDER and MODIS (middle column), for the months of June, July and August 2006 (top row) and for December 2006, January and February 2007 (bottom row). Source: LOA & ICARE*

4

Forward Radiative Transfer in Scattering Atmosphere

It is often necessary to take account of the scattering of radiation by molecules and atmospheric particles (aerosols and clouds), and of its interactions with gas absorption, in order to precisely calculate the radiative quantities. After recapping a few basic concepts concerning atmospheric scattering, in this chapter we present the numerical methods for solving the radiative transfer equation (RTE) used for scattering, absorbing and emitting atmospheric environments. These methods, known as radiative transfer codes, enable us to calculate the radiances and radiative fluxes, as they would be measured by an instrument, supposing a given state of the atmosphere. The objective of this chapter is not to describe these methods in detail, but to present those which are used today, ranging from simple and rapid algorithms to more elaborate models. A list of codes is put forward, along with their main characteristics and domains of application, and their corresponding website. Forward modeling of radiative transfer can be used for remote sensing applications, simulating the radiances measured by a sensor, or for the study of the climate by calculating the radiative fluxes and radiative forcing of the atmosphere. It is also crucial for the development, learning and use of an inverse model which, for its part, is used to retrieve the atmospheric parameters on the basis of a set of radiative measurements (see Chapter 5).

4.1. Atmospheric scattering

Scattering is the spatial redistribution of the light energy of an incident wave when it meets heterogeneity in the propagation medium. It depends on the wavelength of the radiation, on its optical path in the propagation medium, and on the density, nature and size of the scatterers (molecules, clouds or aerosols). The main definitions and characteristics relating to the atmospheric scattering are presented in the following sections.

4.1.1. *Main properties of scattering*

4.1.1.1. *Refractive index*

The refractive index m of an optical medium is a dimensionless factor, which allows us to give a macroscopic description of the propagation of the radiation in that medium. The refractive index is usually complex, such that: $m = n + \mathrm{i}\ \kappa$. The real part, n, is linked to the speed of propagation of light in the material, and the imaginary part, κ, is linked to the attenuation of the light intensity by the material. The index m represents the medium's response to an electromagnetic excitation, and links the material's microphysical properties to its optical properties. The index varies with the wavelength – particularly in the infrared spectrum (Figure 4.1). It also depends on the pressure and the temperature of the medium.

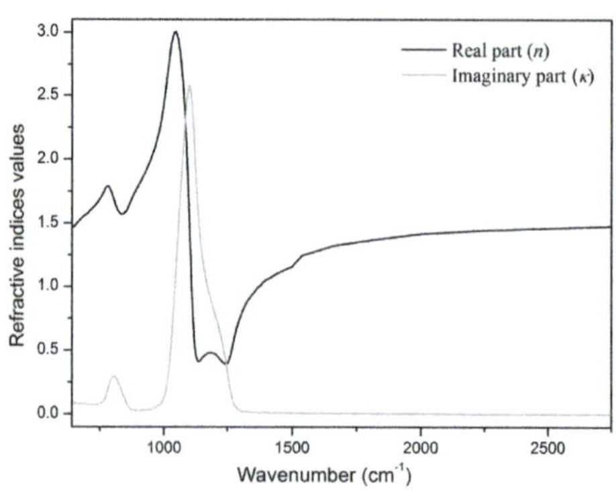

Figure 4.1. *Spectral variation of the real part of the refractive index n (black curve) and imaginary part κ (gray curve) for quartz (SiO_2)*

4.1.1.2. *Size parameter*

The effect of the size of a particle on scattering is generally defined by the size parameter x (no units), which is the ratio of its perimeter to the wavelength of the incident light such that $x = 2\pi r/\lambda$, where r is the radius of the particle (if it is supposed to be spherical) and λ the wavelength of the radiation, both expressed in µm. In particular, the size parameter enables us to define the scattering regime (Figure 4.2):

1) if the size of the particle is much greater than the wavelength in question ($x >> 1$), the scattering regime is indeed described by geometric optics, on the basis of the Snell–Descartes relations;

2) if the size of the particles are near to the wavelength ($x \approx 1$), we need to solve the Maxwell relations, considering the boundary conditions corresponding to the shape of the particles. For spheres, it is Mie scattering, which applies to certain aerosols and water droplets in clouds. The scattered power is inversely proportional to the wavelength, and thus pertains preferentially to longer wavelengths;

3) if the size of the particles is much smaller than the wavelength ($x << 1$), then we have Rayleigh scattering. In the visible and the near-infrared, ultra-fine molecules and particles are involved. In this case, the scattered power is proportional to λ^{-4}, and is therefore greater at shorter wavelengths. This is the reason that the sky is blue.

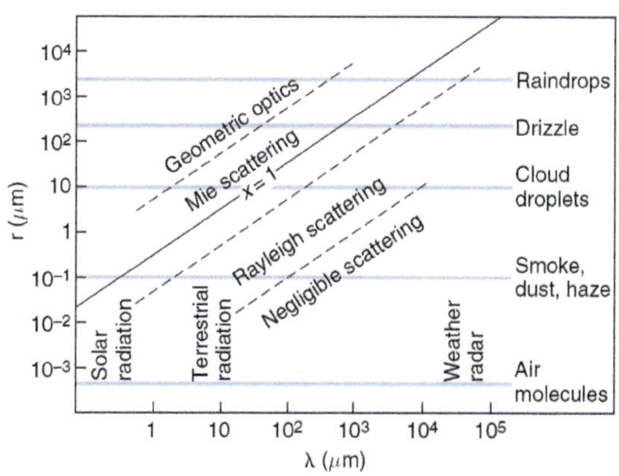

Figure 4.2. *Scattering regime as a function of the wavelength λ and the radius r of the particles, defined by the size parameter x*

4.1.1.3. *Effective cross-sections*

The extinction power P_{ext} [W] is defined as:

$$P_{ext} = P_0 - P_T = P_{scat} + P_{abs} = \sigma_{ext} \cdot E_0, \qquad [4.1]$$

with P_0 being the power of the incident radiation, P_T the power of the transmitted radiation, P_{scat} the power of the scattered radiation, P_{abs} the power of the absorbed radiation, E_0 the illumination of the incident wave [$W.m^{-2}$] and σ_{ext} the effective extinction cross-section [m^2]. We then define the effective cross-section of extinction as being the ratio of the power to the illumination of the incident wave. The effective cross-section expresses the probability of interaction between the radiation and the medium. It depends on the scattering regime and, therefore, on the wavelength of the incident radiation and the optical properties of the medium (size, shape and *m*). From equation [4.1], we obtain the following relation: $\sigma_{ext} = \sigma_{abs} + \sigma_{scat}$. If we take account of the directional aspect, the scattered power in the elementary solid angle $d\Omega$ around the direction (θ, ϕ) is written as: $dP(\theta, \phi) = \sigma(\theta, \phi) I_0 d\Omega$, the angles θ and ϕ respectively being the zenith and azimuthal angles. The differential effective cross-section $\sigma(\theta, \phi)$ is then linked to the effective cross-section of scattering by the relation:

$$\sigma_{scat} = \int_{4\pi} \sigma(\theta, \phi) d\Omega. \qquad [4.2]$$

For spherical or randomly-oriented particles, equation [4.2] is independent of ϕ and becomes: $\sigma_{scat} = 2\pi \int_0^{\pi} \sigma(\theta) \sin\theta d\theta$.

4.1.1.4. *Dimensionless efficiency*

The efficiency factor Q_x is the ratio of the effective cross-section to the surface of the particle projected perpendicularly to the beam. For a spherical particle, this factor is written as $Q_x = 4\sigma_x / \pi D^2$, where D represents the diameter of the particle, the index x refers to the extinction Q_{ext}, the absorption Q_{abs} or the scattering Q_{scat}. As is the case for the effective cross-sections, we have: $Q_{ext} = Q_{abs} + Q_{scat}$.

4.1.1.5. *Single-scattering albedo*

The single-scattering albedo ϖ_0 is defined as the ratio of the scattered part to the extinguished part of the incident light, and is expressed as:

$$\varpi_0 = \frac{\sigma_{scat}}{\sigma_{ext}} = \frac{Q_{scat}}{Q_{ext}} = \frac{Q_{scat}}{Q_{scat} + Q_{abs}}. \qquad [4.3]$$

ϖ_0 takes a value of 1 when only scattering is involved, and 0 in the case of absorption only.

4.1.1.6. *Scattering phase function*

The directional nature of scattering can be described by the scattering phase function $p(\boldsymbol{s}, \boldsymbol{s}')$, which gives the probability that the incident radiation in direction $\boldsymbol{s}$ will be scattered in direction $\boldsymbol{s}'$. For spherical or randomly-oriented particles, the scattering phase function depends only on the angle Θ between the direction of incidence $\boldsymbol{s}$ of the radiation and that of the scattered radiation $\boldsymbol{s}'$. In this case, it is useful to express p as a function of $\cos\Theta \equiv \boldsymbol{s} \bullet \boldsymbol{s}'$. We then define the scattering phase function as:

$$p(\boldsymbol{s}, \boldsymbol{s}') = \frac{4\pi}{\sigma_{scat}(\cos\Theta)} \frac{d\sigma_{scat}(\cos\Theta)}{d\Omega} = \frac{4\pi\sigma}{\sigma_{scat}(\cos\Theta)},$$

with σ representing the differential scattering effective cross-section. It is typically normalized to 4π:

$$\frac{1}{4\pi}\int_{4\pi} p\ (\cos\Theta)\, d\Omega = \frac{1}{4\pi}\int_0^{2\pi}\int_0^{\pi} p\ (\cos\Theta)\ \sin\Theta d\Theta df = 1 \qquad [4.4]$$

or $\int_{-1}^{1} p(\cos\Theta)\, d\cos\Theta = 2.$

4.1.1.7. *Asymmetry factor*

The asymmetry factor g tells us about the quantity of light scattered forward and backward:

$$g = \frac{1}{4\pi}\int_{4\pi} p(\cos\Theta)\cos\Theta d\Omega$$

and therefore

$$g = \frac{1}{2}\int_{-1}^{1} p(\cos\Theta)\cos\Theta d\cos\Theta \qquad [4.5]$$

For spherical or randomly-oriented scatterers, the value of g lies between -1 and 1. That parameter has the advantage of summarily describing the angular distribution of the scattered radiation. Indeed, a positive value indicates that the scattering in the direction of the incident wave will be predominant. Conversely, a negative value shows that the light will be scattered primarily in the opposite direction (back scattering). Finally, a value of g near to zero corresponds to symmetrical or isotropic scattering.

4.1.2. *Rayleigh scattering*

When it interacts with the material, the incident electromagnetic wave, because of its electrical field, creates an electrostatic dipole which emits induced radiation in all directions. In 1873, John William Strutt (better known as Lord Rayleigh) showed that such an electric dipole radiates a quasi-isotropic electromagnetic field, whose phase function is expressed as $p(\Theta) = 3/4(1+\cos^2\Theta)$. The distribution of the radiation, therefore, is symmetrical (Figure 4.3), implying that the asymmetry coefficient g is null. In this case, the differential effective cross-section of scattering is written as:

$$\sigma(\Theta) = \frac{\pi^4 D^6}{8\lambda^4}\left(\frac{m-1}{m+2}\right)^2\left(1+\cos^2\Theta\right) \qquad [4.6]$$

where Θ is the scattering angle, m and D are respectively the complex refractive index and the diameter of the particle. The Rayleigh scattering efficiency and absorption factors are such that:

$$Q_{scat} = \frac{8}{3}\alpha^4 \left| \frac{m^2 - 1}{m^2 + 2} \right|^2 \text{ and } Q_{abs} = 4\alpha \,\mathrm{Im}\left(\frac{m^2 - 1}{m^2 + 2} \right). \qquad [4.7]$$

Moreover, the intensity of the scattered radiation is proportional to λ^{-4}. Thus, Rayleigh scattering favors the shorter wavelengths in the spectrum. With incident white light, therefore, blue ($\lambda \sim 400$ nm) is scattered much more intensely than red ($\lambda \sim 700$ nm). This is what accounts for the blue color of the sky, where the scattering centers are the molecules present in a high concentration in the atmosphere (primarily O_2 and N_2), and the very fine particles. At dawn or at dusk, however, when the Sun is low on the horizon, the atmospheric layer traversed by the radiation is much thicker than it is during the day. Then, short wavelengths are totally scattered and we then only see the longest wavelengths (red), with the sky appearing reddish-orange in the direction of the Sun.

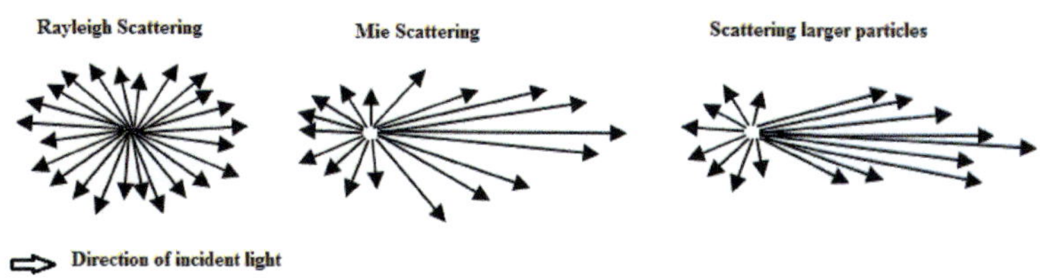

Figure 4.3. *Illustration of the scattered radiation field in the case of Rayleigh scattering (left), Mie scattering in the case of small particles (middle) and Mie scattering for large particles (right). The incident wave is coming from the left*

4.1.3. *Mie scattering*

Mie's solution (1908) is an analytical solution to the Maxwell relations describing the interaction of light with spherical scatterers for which $x \approx 1$. The solution presents itself in the form of infinite sums of spherical harmonics [BOR 99; VAN 81]. For $x \ll 1$, it again gives us the Rayleigh scattering. According to Mie's theory, the amount of energy scattered forward is greater than in the other directions, and increases with the size of the scatterer (Figure 4.3). Moreover, it is less dependent on the wavelength than in the case of Rayleigh scattering. Consequently, as the droplets in a cloud are very large in comparison to the wavelength of visible light, the scattering regime in this case is Mie scattering, which is uniform across all the colors in the spectrum (hence the white color of the clouds), but is nonetheless anisotropic particularly in the forward direction. Finally, the intensity of the scattered light increases with the difference between the refractive indices of the scattering centers and the surrounding medium. Thus, for instance, modern white paints contain mineral species chosen for their very high refractive index, such as cinnabar ($m > 3$), compared to the index of the binder, such as oil ($m \approx 1.5$), for example. Hence, these scatterers are the site of significant Mie scattering, producing a brilliant white color. Clouds, however, are of a less intense white, because the difference between the refractive index of the water droplets ($m \approx 1.3$) and the air surrounding them ($m = 1$) is smaller.

4.1.4. *Non-spherical particles*

For scatterers which exhibit low dissymmetry and/or which are oriented randomly, the scattering is averaged and the gap with the results obtained by Mie's theory is then minimal [PUJ 12]. Max Born also showed that there are exact analytical solutions to the Maxwell relations for cylinders and ellipsoids [BOR 99]. However, the solutions provided by Mie scattering have no analytical expression for more complex shapes, and we are therefore limited to numerical approaches. Thus, when the particles are ellipsoidal in shape (i.e. they are flattened or elongated), the T-matrix theory [MIS 96] can be used for a rather effective and rapid calculation. When the particles are of complex or random shapes, other theories can be used [MIS 00], such as the finite element method [SIL 96, YAN 98]. These techniques are highly accurate, but are also very costly in terms of computation time, which is their main drawback. Numerical codes to compute Mie scattering, or based on the T-matrix, are given in Table A.4 in the Appendix.

4.1.5. *Extinction coefficient and optical thickness*

The extinction coefficient k_{ext} is the proportion of energy lost by the wave, by absorption or scattering, per unit length, when passing through a medium:

$$k_{ext}(\lambda)=\int \pi r^2 Q_{ext}(r,\lambda,m)n(r)dr \quad [4.8]$$

where r is the mean geometric radius, $n(r)$ is the size distribution (see Chapter 1) and Q_{ext} is the extinction efficiency factor. Figure 4.4 shows the spectral variation in the infrared of k_{ext} for water drops, depending on their size. It is important to note that this spectral variability stems from the dependence on scattering, and therefore the complex refractive index m, at the wavelength in question. This spectral signature is at the heart of the sensitivity of infrared measurements to the type of particles and the particle size distribution.

One of the parameters most frequently used for studying atmospheric particles by remote sensing is the optical thickness τ_λ, because its value contains information both about the number of particles along the optical path and about their extinction properties. Its expression is derived from the Beer–Lambert law: for a medium whose thickness is L, we have $\tau_\lambda=-ln(I_\lambda/I_{0\lambda})$, where $I_{0\lambda}$ is the intensity of the incident radiation and I_λ is the intensity at distance L. Hence, we can define the optical thickness for an optical path between two points s_1 and s_2 as:

$$\tau_\lambda(s_1,s_2)=\int_{s_1}^{s_2} k_{ext}(\lambda,s)ds\,. \quad [4.9]$$

For a homogeneous medium (i.e. k_{ext} constant), this relation gives us $\tau_\lambda = k_{ext}.(\lambda).L$. The optical thickness implicitly contains the processes of scattering and absorption: $\tau_\lambda = \tau_{\lambda,abs} + \tau_{\lambda,scat} = \tau_{\lambda,ext}$, where $\tau_{\lambda,abs}$, $\tau_{\lambda,scat}$, and $\tau_{\lambda,ext}$, are respectively the optical thicknesses of absorption, scattering and extinction.

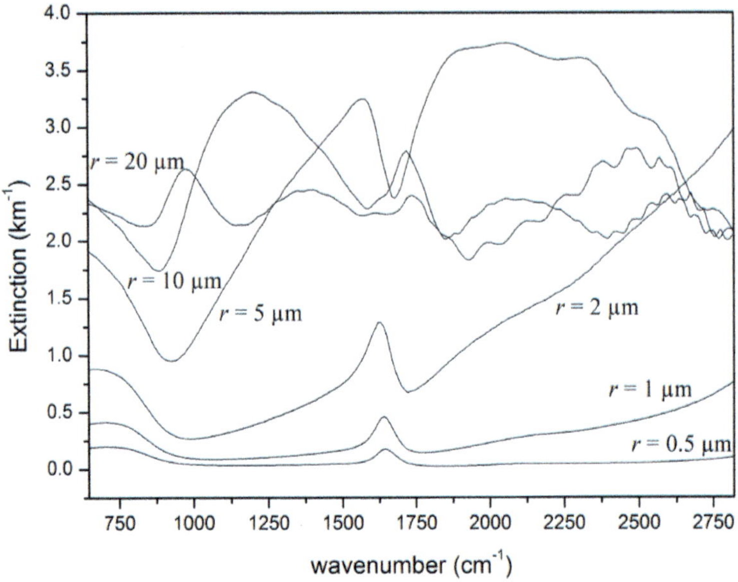

Figure 4.4. *Spectral variation of the extinction coefficient k_{ext} of water drops as a function of the mean geometric radius r of the size distribution*

4.2. Polarization

Natural sunlight is not polarized, but it becomes polarized by scattering or reflection by a terrestrial surface, a cloud or an aerosol plume. In spatial remote sensing, this phenomenon is well known, and has been being exploited for the past twenty years by the instrument polarization and directionality of the Earth's reflectances (POLDER). POLDER takes multi-angle polarized measurements, in the visible and near-infrared areas of the spectrum, dedicated to clouds and aerosols. From 2020, POLDER measurements will be continued with the 3MI instrument (multi-viewing multi-channel multi-polarization imaging), which will have 10 channels polarized between 0.41 and 2.13 μm, and will be aboard the MetOp-SG satellite. Moreover, several recent missions devoted to infrared measurements, such as Orbiting Carbon Observatory-2 (OCO-2, NASA, USA) and greenhouse gases observing satellite (GOSAT, JAXA, Japan), have also allowed us to observe the Earth's atmosphere in polarization. All of these missions have demonstrated that this technical specificity constitutes an interesting way for better determination of the optical and microphysical parameters of the atmospheric particles.

Polarization, by definition, corresponds to the direction of the electrical field $\boldsymbol{E}$ of the electromagnetic radiation. If the direction of the field $\boldsymbol{E}$ is random over time and equiprobable in its orientation, the radiation is said to be non-polarized. In the case of a monochromatic plane wave, the general expression of the electrical field $\boldsymbol{E}$ of the wave is of the form:

$$\boldsymbol{E} = E_{0x}\cos\left(\omega t - \boldsymbol{k}\cdot\boldsymbol{r}\right)\boldsymbol{e}_x + E_{0y}\cos\left(\omega t - \boldsymbol{k}\cdot\boldsymbol{r} - \Phi\right)\boldsymbol{e}_y, \quad [4.10]$$

where $\omega = 2\pi c/\lambda$ is the pulsation of the wave, $\boldsymbol{r}$ the position vector and $\boldsymbol{k}$ the wave vector. We can distinguish the following states of polarization, illustrated in Figure 4.5:

1) linear polarization: $\boldsymbol{E}$ maintains a specific direction throughout its propagation, and $\Phi = 0$;

2) circular polarization: the edge of the electrical field forms a circle travelled in the trigonometric sense (left polarization) or in the opposite direction (right polarization), for an observer receiving the wave ($E_{0x} = E_{0y}$ and $\Phi = \pm\,\pi/2$);

3) elliptical polarization: the edge of the field $\boldsymbol{E}$ forms an ellipse over time. This is the most common situation;

4) if the phase difference Φ is a random function of time, then the electrical field has a random direction and the light is not polarized.

These elementary polarizations serve as a basis for the decomposition of the field. The polarization of a medium stems from the overall effect of all of the scatterers. Thus, the total polarization is the sum of the elementary polarizations.

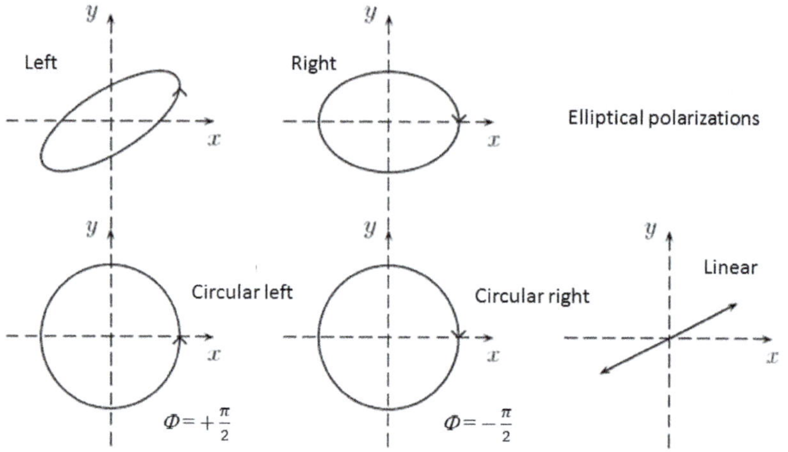

Figure 4.5. *Illustration of the different types of polarization*

A polarized wave, therefore, is characterized by three parameters: the amplitudes E_{0x} and E_{0y} and the phase difference Φ. In optics, the most usual representation is that put forward by Stokes (1852). It links together the Cartesian coordinates, Q, U and V, of the point representative of the polarized wave of intensity I on the Poincaré sphere with the triplet (E_{0x}, E_{0y} and Φ). The advantage of the Stokes representation is to offer a simple representation of the polarization of a wave by a column matrix whose four rows are I and the Stokes parameters Q, U and V (Table 4.1). The first element I corresponds to the total scalar radiance, while the elements Q, U and V provide information on the degree of polarization. In practice, the actual light consists of a random mixture of simple waves in very rapid succession, and it is never ideally coherent. In this case, incoherent radiation is unpolarized or partially polarized. The degree of polarization is then defined as $p = \sqrt{Q^2 + U^2 + V^2} / I$, with $Q = U = V = 0$ for completely unpolarized radiation. However, ideally-coherent radiation is fully polarized and the intensity is then defined as $I^2 = Q^2 + U^2 + V^2$.

Stokes parameters	Unpolarized	Linearly polarized				Circularly polarized	
		0°	90°	45°	-45°	left	right
I	1	1	1	1	1	1	1
Q	0	1	-1	0	0	0	0
U	0	0	0	1	-1	0	0
V	0	0	0	0	0	-1	1

Table 4.1. *Stokes parameters for the usual states of polarization*

We can express the parallel and perpendicular components of the scattered electrical field $\boldsymbol{E}_{scat}$ as a function of the incident field $\boldsymbol{E}_{inc}$, by relation $\mathbf{E}_{scat} = \mathbf{S} \cdot \mathbf{E}_{inc}$, where **S** represents the matrix of the scattering amplitudes, generally called the Mueller matrix, which depends on the scatterer (shape, size and nature). For a scatterer of any shape, we write:

$$\begin{bmatrix} I_{scat} \\ Q_{scat} \\ U_{scat} \\ V_{scat} \end{bmatrix} = \frac{1}{k^2 r^2} \begin{bmatrix} S_{11} & S_{12} & S_{13} & S_{14} \\ S_{21} & S_{22} & S_{23} & S_{24} \\ S_{31} & S_{32} & S_{33} & S_{34} \\ S_{41} & S_{42} & S_{43} & S_{44} \end{bmatrix} \begin{bmatrix} I_{inc} \\ Q_{inc} \\ U_{inc} \\ V_{inc} \end{bmatrix}, \qquad [4.11]$$

where k is the wavenumber (ω / c), at a distance r from the scatterer. For example, Van de Hulst (1957) showed that for a set of scatterers with simple symmetry (sphere, cylinder and ellipsoid) oriented randomly, certain elements of the matrix **S** are null:

$$\begin{bmatrix} I_{scat} \\ Q_{scat} \\ U_{scat} \\ V_{scat} \end{bmatrix} = \frac{1}{k^2 r^2} \begin{bmatrix} S_{11} & S_{12} & 0 & 0 \\ S_{12} & S_{22} & 0 & 0 \\ 0 & 0 & S_{33} & S_{34} \\ 0 & 0 & -S_{34} & S_{44} \end{bmatrix} \begin{bmatrix} I_{inc} \\ Q_{inc} \\ U_{inc} \\ V_{inc} \end{bmatrix}. \qquad [4.12]$$

In addition, for spheres: $S_{11}=S_{22}$ and $S_{33}=S_{44}$.

From an experimental standpoint, polarized light can be analyzed by determining the values Q, U and V on the basis of measurements of intensity, using linear polarizers which select a preferential direction of polarization. Polarized light is transmitted if the axis of transmission of the polarizer is parallel to its polarization; it is totally absorbed by the polarizer if that axis is perpendicular.

4.3. Radiative transfer equation (RTE) in a scattering medium

4.3.1. *General expression of the RTE*

The purpose of the RTE is to describe the propagation of the radiation in the atmospheric medium, taking account of the phenomena of absorption, scattering and heat emission. It is an equation of transport of monochromatic radiance L_λ, which expresses the conservation of radiative energy in an elementary volume, for incident radiation in the direction of propagation $\boldsymbol{s}$ (Figure 4.6). It is helpful to remember that the calculation of the directional radiance, by way of the RTE, stems from geometric optics. This approach is valid because the interferences and/or coherence between the scatterers can be ignored in the atmosphere. If not, it becomes necessary to deal with the fields $\boldsymbol{E}$, $\boldsymbol{B}$ and the Poynting vector. We use the symbol s for the curvilinear abscissa along the direction of propagation. The variation of the spectral radiance along the elementary path ds is written as:

$$dL_\lambda(\mathbf{s}) = -k_{ext,\lambda} L_\lambda(\mathbf{s})\,ds + k_{ext,\lambda} J_\lambda(\mathbf{s})\,ds\,. \qquad [4.13]$$

The first term on the right-hand side expresses the attenuation of the radiance, which is proportional to the extinction coefficient $k_{ext,\lambda}$. The second term represents the increase of the radiance along the path. We define J_λ as the source function, which has the dimension of a radiance and is expressed in the same units. It is the sum of the source functions due to multiple scattering J^S and thermal emission J^E:

$$J_\lambda(s) = J_\lambda^S(s) + J_\lambda^E(s) = \frac{\varpi_{o,\lambda}}{4\pi}\int_{4\pi} p_\lambda(s,s')L_\lambda(s')d\Omega' + [1-\varpi_{o,\lambda}]B_\lambda(T) \qquad [4.14]$$

where $\varpi_{0\lambda}$ is the single-scattering albedo, p_λ the phase function, B_λ the Planck function and **s**' the scattering direction. For reasons of simplicity, hereinafter we will omit the spectral indicator λ. Conventionally, the RTE is often expressed as a function in spherical coordinates, on the basis of the zenith and azimuth angles (θ, ϕ), the air mass factor $\mu = cos\theta$ and the optical thickness τ of the medium, such that $d\tau = -\mu\sigma_{ext}ds$. Note that, by convention, $\mu > 0$ corresponds to upward radiation and $\mu < 0$ to downward radiation. It is also easier to separate the radiation field into two components such that: $L(\tau,\mu,\varphi) = L_s(\tau,\mu,\varphi) + L_d(\tau,\mu,\varphi)$. The first is the solar or direct component L_s, which represents the part of the solar radiation which has not been scattered:

$$L_s(\tau,\mu,\varphi) = \delta(\mu-\mu_0)\delta(\varphi-\varphi_0)F_0 \exp\left(-\frac{\tau}{\mu_0}\right), \qquad [4.15]$$

where F_0 is the solar function, (μ_0, ϕ_0) the direction of solar incidence and δ the Dirac delta function. The second is the diffuse component L_d, which represents the radiation having been scattered at least once. By simplifying the notations, the RTE can be expressed, for the diffuse radiance, in scalar form, as follows:

$$\mu\frac{dL}{d\tau}(\tau,\mu,\phi) = L(\tau,\mu,\phi) - \frac{\varpi_o}{4\pi}\int_0^{2\pi} d\phi' \int_{-1}^{1} d\mu' p(\mu',\phi',\mu,\phi)L(\tau,\mu',\phi')$$
$$-[1-\varpi_o]B(T) - \frac{\varpi_o}{4\pi}p(\mu_0,\phi_0,\mu,\phi)F_0 \exp\left(-\frac{\tau}{\mu_0}\right) \qquad [4.16]$$

If we take account of the polarization of light, this equation is expressed in vectorial form, with the Mueller matrix ***S*** and the scattering matrix ***p***. The radiance ***L*** is then represented by its four Stokes components (*L, Q, U* and *V*), with the first component being calculated by means of the RTE in scalar form.

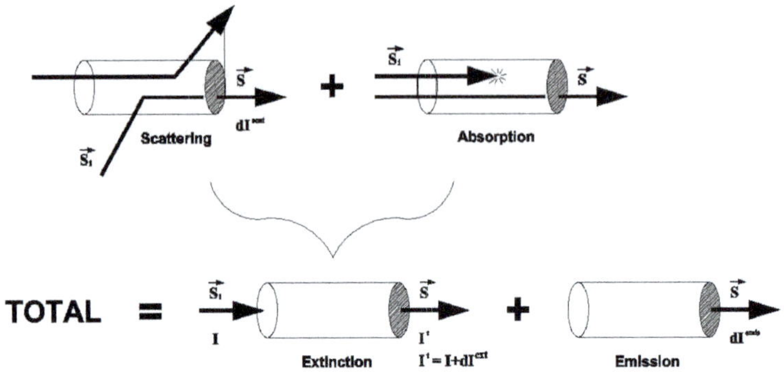

Figure 4.6. *Radiative energy balance in a volume element illustrated by the extinction (1), the scattering (2) and the emission (3) of the radiation in direction* ***s***

4.3.2. *Solving of the RTE*

The precise solution of the RTE, taking account of all the radiative processes, is a complicated problem because the spatial distribution of the atmospheric components is very heterogeneous, and the boundary conditions are complex. Moreover, the problem involves five variables: three spatial variables (x, y and z) and the two angular variables (θ and ϕ) to describe the radiative quantities. There is not exact analytical solution and numerical methods are necessary. We can distinguish statistical methods, such as the Monte Carlo approach, which calculate the desired radiative value for any geometry, and explicit methods, known as "plane–parallel" methods, which solve the RTE for the whole of the radiation field, discretized for calculation in a certain number of directions.

4.3.2.1. *"Three-dimensional" (3D) statistical methods*

This approach his the most accurate, because it takes account of the spatial heterogeneities. The atmosphere is divided into homogeneous voxels (the word is a contraction of "volume" and "pixel"), of dimensions dx, dy and dz. The radiative quantities are then calculated using a method known as the Monte Carlo method, which consists of simulating the trajectories of the fictive light particles (FLIPs – see the

Preface) using random procedures. Thus, the precision depends on the number of trajectories simulated and, therefore, it is directly linked to the computation time. However, this approach uses statistical methods, which are capable of handling a very large number of FLIPs in reasonable computation times. The power of today's computers and the development of new numerical techniques, such as parallel computation or graphic cards, also make it easier to use these codes. In addition, the Monte Carlo simulation has the advantage of considering the exact phase function $p(\Theta)$ of the scatterers, which is identified as the probability of scattering defined by the relative scattering angle Θ.

We can distinguish two types of Monte Carlo codes. Forward codes simulate radiative transfer from the source to the detector for a given angle or in a solid angle. This approach requires significant computation times if the probability of the radiation reaching the detector is slight. It is therefore more advantageous to use a backward code, based on the principle of reciprocity of light. The trajectories of the FLIPs follow an inverse path from the detector, for a given viewing angle, and they are reconstructed at the source in the propagation medium as a function of its optical properties. This approach is well-suited for broad radiative sources and small detectors, because it avoids generating optical trajectories leading from the source but which never reach the detector. These methods will not be detailed in this book, but readers will find a non-exhaustive list on the website of the Intercomparison of 3D Radiation Codes (I3RC): http://i3rc.gsfc.nasa.gov, and a state of the art on 3D radiation codes in [MAR 05].

4.3.2.2. *Explicit "plane–parallel" methods*

The atmospheric medium is stratified into homogeneous layers, horizontally infinite, plane and parallel (Figure 4.7). Each layer is defined by the temperature and pressure at its boundaries, and the optical parameters (optical thickness, single-scattering albedo and phase function). The solution to that equation requires us to solve a system of integro-differential equations with boundary conditions (i.e. the properties of the surface and solar radiation from space). This simplified representation of the atmosphere, often spoken of as a 1D representation, enables us to solve the RTE quickly. Therefore, it is often used for applications requiring a large number of simulations, such as climate modeling or operational treatment of satellite data.

The usage of these methods for spatially-extensive scenes, such as satellite images, means that we need to discretize the atmosphere into independent atmospheric columns, each of which are divided into boxes of varying vertical resolution – often finer in the troposphere, where the majority of radiative processes take place. The spatial resolution of the surface of those columns defines the size of the corresponding pixel (see Chapter 2). The three-dimensional (3D) effects of the radiative transfer in a heterogeneous medium, such as horizontal exchanges of radiation from one column to another (or from one pixel to another), are not taken into account (independent column approximation/independent pixel approximation – ICA/IPA). A second source of error is the difference between the layers (or pixels), which are supposed to be homogeneous in the plane-parallel approximation, and those which are actually observed, and are 3D and heterogeneous (plane–parallel approximation–PPA). Figure 4.8 illustrates the estimation of the error committed as regard the radiance simulated at the peak of the atmosphere using a "plane–parallel" method, as a function of the size of the pixel. This error is greater with small pixels (effect of the ICA/IPA) or large pixels (effect of the PPA). It is minimal for pixels of around 1 km size. Note, in this regard, that the majority of spatial remote sensing instruments have a spatial resolution of between 500 m and 15 km. This range corresponds to the minimum error committed by a 1D-type approach.

The "plane–parallel" approximation may also be too approximate for solar zenith angles of close to 90° (at sunrise or sunset, geometries of sight to the limb). A correction consisting of slightly modifying the viewing geometry is therefore used to take account of the sphericity of the atmospheric layers; it also enables us to take account of the refractivity of the atmospheric layers [GOR 94].

The main methods and codes used to solve the RTE in the hypothetical case of a 1D atmosphere divided into homogeneous, plane and parallel layers are presented in the following sections.

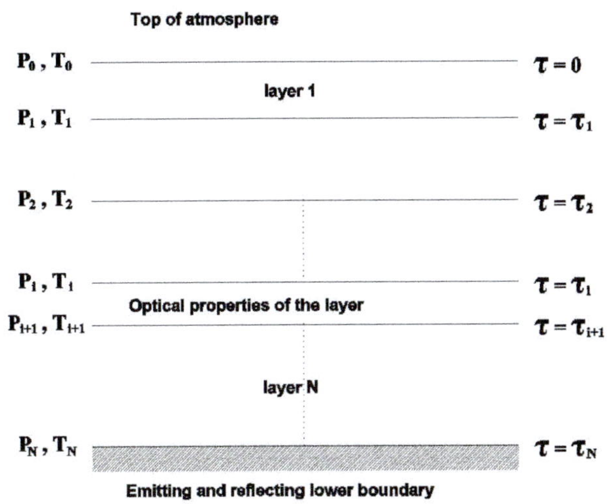

Figure 4.7. *Diagrammatic representation of an atmosphere divided into N homogeneous, plane and parallel layers*

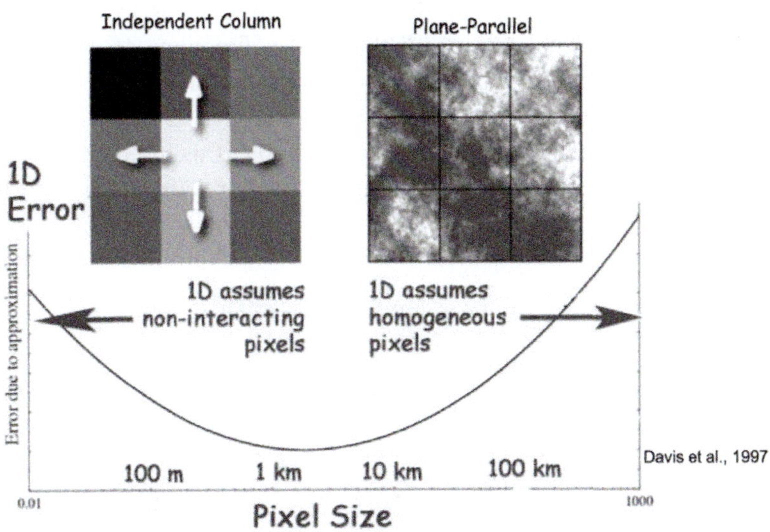

Figure 4.8. *Estimation of the error committed by a 1D model of the radiative transfer in relation to the 3D reference calculations performed with a Monte Carlo code, as a function of the size of the pixel in question. Source: A. B. Davis*

4.3.3. *Azimuthal dependence of the radiation field*

Beyond Rayleigh scattering, scatterers generally have a complex scattering phase function $p(\cos\Theta)$ which is not expressed in a simple mathematical form (Figure 4.9). We then use approximations in order to facilitate numerical solving of the RTE. A conventional approach consists of expressing $p(\cos\Theta)$ as an infinite Legendre series $P_l(\cos\Theta)$. In practice, this sum is reduced to a number N of terms:

$$p(\cos\Theta) = \sum_{l=0}^{\infty} \beta_l P_l \cos\Theta \approx \sum_{l=0}^{N-1} \beta_l P_l \cos\Theta, \qquad [4.17]$$

with β_l being the coefficients of the Legendre polynomial expansion defined as:

$$\beta_l = \frac{2l+1}{2} \int_{-1}^{1} P_l(\cos\Theta) p(\cos\Theta) d(\cos\Theta). \qquad [4.18]$$

The first coefficient β_0 is always equal to 1, so as to respect the conditions of normalization of the phase function. If the phase function is isotropic, all the coefficients have a value of 0 except for β_0. The asymmetry factor g, for its part, is defined as $\beta_1/3$. Note that we often define the phase function on the basis of its moments $m_l = \beta_l/(2l+1)$. Similarly, it is common to use a Fourier series expansion for the radiance:

$$L(\tau, \mu, \varphi) \approx \sum_{m=0}^{2N-1} L^m(\tau, \mu) \cos[m(\varphi_0 - \varphi)]. \qquad [4.19]$$

By substituting equations [4.17] and [4.19] into the general RTE [4.16], we obtain an equation for each term of the Fourier-series expansion. These numerical expansions allow us to "eliminate" the azimuthal dependence on the RTE, the Fourier terms m being decoupled. The solutions obtained for each term m enable us to obtain the azimuthal components L^m, the sum of which is defined by equation [4.19] and gives the complete azimuthal dependence of the intensity [THO 02].

4.3.4. *Simplification of the phase function*

The Legendre polynomial expansion allows us to accurately model anisotropic phase functions if N is large, which requires significant computation times, especially in solar infrared (e.g. for ice crystals). A very common approximation in radiative transfer consists of truncating the phase function so as to reduce the number of terms N that are needed to recover it, without adversely affecting the precision of the calculated radiative quantities. This truncation consists of replacing the forward peak of the phase function with a Dirac delta function:

$$p(\cos\Theta) \approx 2f\delta(1-\cos\Theta) + (1-f)\,p'(\cos\Theta) \qquad [4.20]$$

where $p'(\cos\Theta)$ is the approximate phase function and f is the truncation factor. This approximation entails considering that the energy scattered in the forward peak of the phase function is directly transmitted forward ($\Theta = 0$). Various truncations can be used, such as the δ-Potter, the δ-M or the δ-fit [COM 12]. The initial medium is then replaced by an optically-equivalent medium. For example, for the δ-fit, the equivalent optical properties (m_l^*, ϖ_0^* and σ_e^*) are expressed on the basis of the initial properties (m_l, ϖ_0 and σ_e) as:

$$m_l^{eq} = \frac{m_l - f}{1-f}, l = 0,\dots,2N-1,\ \varpi_0^{eq} = \frac{\varpi_0(1-f)}{1-\varpi_0 f};$$
$$\sigma_e^{eq} = (1-\varpi_0 f)\sigma_e. \qquad [4.21]$$

These truncations yield very good results for the simulation of the flux, but can lead to significant errors with regard to the radiance – particularly with polarized light. The use of a single-scattering correction (TMS correction) then very greatly improves the results [COM 12]. In the case of droplets in a cloud, Figure 4.9 shows that the δ-fit truncation enables us to reconstruct the nominal scattering function with a very high degree of accuracy on the basis of a reduced number N. On the other hand, the use of the δ-M leads to oscillations on the phase function obtained after truncation.

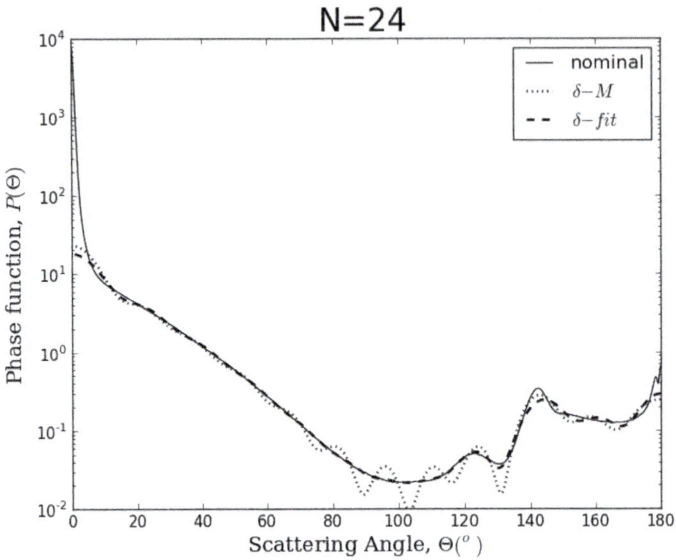

Figure 4.9. *Nominal phase function p(Θ) at 412 nm (solid line) and calculated on the basis of a Legendre polynomial development and the truncation by the method of the δ-M or the δ-fit, where N = 24 and for a model of cloud droplets defined by [KOK 10]*

Similarly, the Henyey–Greenstein function (1941) is a simple analytical expression which best represents the angular dependence of scattering $p(\Theta)$ by small particles (Figure 4.10). Its normalized scatter phase function is expressed in terms of g:

$$p(\Theta)=\frac{1}{4\pi}\frac{1-g^2}{\left(1+g^2-2g\cos\Theta\right)^{3/2}}. \quad [4.22]$$

The moments of the phase function are defined as $m_l = g^l$. This function is widely used because it enables us to define a realistic phase function on the basis solely of the asymmetry factor and using a reduced number N of moments.

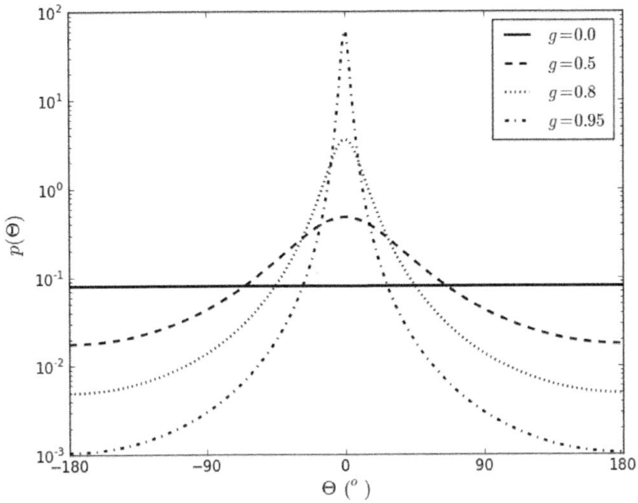

Figure 4.10. *Henyey–Greenstein function p(Θ) for different values of the asymmetry factor g*

4.4. Numerical methods to solve the RTE in a scattering plane–parallel medium

In this section, we briefly describe a few typical methods. At the end of this chapter, readers will find a number of references giving more in-depth information about each of these methods. As scattering is greater at short wavelengths, these methods have generally been developed for the solar domain, and then extended to the infrared. Indeed, for a long time, scattering was overlooked in the field of infrared radiation. However, recent studies have demonstrated that this effect may be significant – e.g. with mineral aerosols [DUF 02] or clouds of ice [DUB 08]. These methods for solving the RTE are then used to calculate the radiative quantities (radiance, radiative flux, etc.), on the basis of the parameters characterizing the atmospheric medium (gases, aerosols and clouds) and the surface (Figure 4.11). In the case of a heterogeneous atmosphere divided into plane, parallel and homogeneous layers, the average value of each of these input parameters needs to be defined for each atmospheric layer.

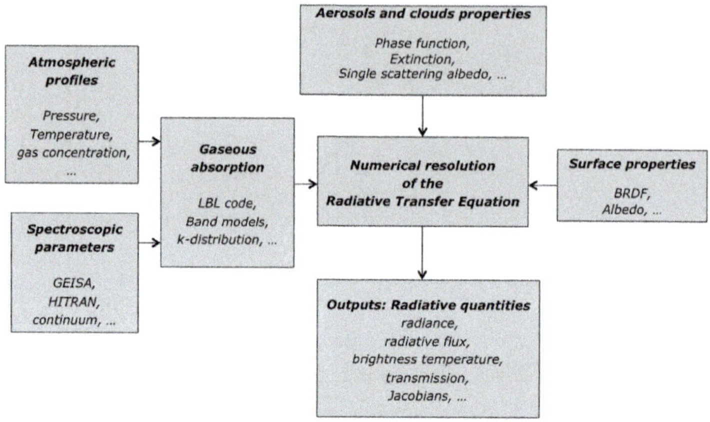

Figure 4.11. *Diagrammatic representation of the solving of the radiative transfer equation*

4.4.1. *Approximate analytical expressions*

Approximate analytical formulations have been developed to solve the RTE. However, they do not give us a precise solution for heterogeneous atmospheres. However, we can point to two simple and useful formulations:

1) The primary scattering approximation enables us to calculate the radiance of a medium for which multiple scattering is supposed to be negligible, which means that the average distance between the scatterers is much greater than the size of those scatterers, and therefore, the optical thickness is slight. The radiance L^{TOA} emerging at the top of the atmospheric layer ($\mu > 0$) can be approximated by:

$$L^{TOA}(\mu,\varphi) \approx \frac{\varpi_0}{4\pi} F_0 p(\mu,\varphi,\mu_0,\varphi_0) \frac{\mu_0}{(\mu_0+\mu)}\left(1-\exp\left[-\left\{\frac{1}{\mu}+\frac{1}{\mu_0}\right\}\tau\right]\right)+(1-\varpi_0)B\left(1-\exp\left[-\frac{\tau}{\mu}\right]\right) \quad [4.23]$$

This expression can be simplified, in the case of primary scattering ($\tau \ll 1$), as:

$$L^{TOA}(\mu,\varphi) \approx \frac{\varpi_0}{4\pi} F_0 p(\mu,\varphi,\mu_0,\varphi_0)\frac{\tau}{\mu}+(1-\varpi_0)B\left(1-\exp\left[-\frac{\tau}{\mu}\right]\right). \quad [4.24]$$

This formulation enables us to estimate the radiance on the basis of a simple analytical expression, which can serve as a starting point for a more complex method such as those described in the next sections.

2) For azimuth-independent cases, another simple approach is to suppose that the radiation can only propagate in two discrete directions. As the radiative fluxes are azimuthally averaged quantities, we can expect that the angular variations of the radiation field will not have much influence over the calculation of those quantities. This approach, which is known as the "Two Stream Method", is therefore unable to provide a good degree of accuracy as regards the directional radiances, but on the other hand it performs far better for the simulation of average hemispherical fluxes. It is still very widely used – particularly in general circulation or weather forecasting models, for which simplicity and rapidity are important assets. For example, codes are available at the addresses below[1].

4.4.2. *Discrete ordinate method*

The discrete ordinate method, put forward by Chandrasekhar in the 1950s, consists of approximating the integral term in equation [4.16] by a finite sum, on the basis of a double Gauss quadrature (one for each hemisphere) in the interval [−1, + 1]:

$$\int_{-1}^{+1} f(\mu) d\mu = \sum_{\substack{i=-N \\ i \neq 0}}^{N} a_i f(\mu_i), \qquad [4.25]$$

where μ_j and a_j are, respectively, the points and the weights of quadrature such that ($\mu_{-j} = -\mu_j$) and ($a_j = a_{-j}$). The radiance is then calculated for a finite number of directions μ_j, associated with a Fourier-series expansion for the radiance and the scattering phase function.

1 http://fapar.jrc.ec.europa.eu/WWW/Data/Pages/FAPAR_Software/FAPAR_Software_RTModels_two-stream.php?blnJS=0 or http://www.rtslidort.com/otherprod.html.

The numerical software tool DISORT (Table 4.2) can solve the RTE by this method [STA 88]. This program is very widely used in the community, and is able to calculate the radiances and monochromatic radiative fluxes across the whole of the electromagnetic spectrum, for absorbing and scattering media, including thermal emission and without limitation for gases, aerosols and clouds. Let us also note the existence of the Spherical Harmonic Method, based on the separation of the azimuthal dependence and the use of an expansion of the radiance into Legendre polynomials, which is conceptually very similar to the discrete ordinate method. Primarily, it differs from it by the numerical treatment of the boundary conditions. We can cite the example of the SHDOM code (spherical harmonics discrete ordinate method for 3D atmospheric radiative transfer) which uses both the discrete ordinate method and the spherical harmonic method simultaneously to solve the RTE for 3D atmospheres.

4.4.3. *Adding-doubling method*

The basic principle of this method is relatively simple: if the functions of reflection and transmission of two scattering and absorbing layers are known, those same functions of reflection and transmission for the two combined layers can be obtained by calculating the successive reflections between the two layers. If, in addition, those two layers are chosen as being homogeneous and identical, the calculation for a thick layer can be obtained quickly (principle of "doubling"). It is practical to begin this procedure by supposing we have very thin layers, facilitating the use of the single-scattering approximation. With a homogeneous layer, this process is then repeated until we obtain the total desired optical thickness. The term "adding", for its part, refers to the use of the functions of reflection and transmission for the calculation of the solutions in the case of combinations of layers with different optical properties [DEH 87, THO 02].

Numerical programs have been developed on the basis of the adding-doubling method to solve the RTE in matrix form, for atmospheres including gases, aerosols and clouds, taking account of the polarization of the radiation. The concept of adding-doubling is fairly similar to that of the matrix operator method, which employs the interaction principle to calculate the transmission and reflection functions, instead of infinite sums, as in the case of the adding-doubling method.

4.4.4. *Successive orders of scattering method*

In section 4.4.1, we have seen that it is possible to analytically solve radiative transfer in the case of primary scattering. In the general case of scattering media, a more accurate solution is to develop the source function source in line with the successive scatterings [LEN 07]. The source function of the FLIPs that undergo scattering of order n is then written as:

$$J^n(\tau,\mu,\varphi)=\frac{\varpi_0}{4\pi}\int_{-1}^{1}\int_{0}^{2\pi}p(\mu,\varphi,\mu',\varphi')L^{(n-1)}(\tau,\mu',\varphi')d\mu' d\varphi' \qquad [4.26]$$

where $L^{(n-1)}(\tau,\mu',\phi')$ corresponds to the FLIPs which have undergone $(n-1)$ scatterings. The order 0 corresponds to direct radiation, which has not undergone scattering. Hence, this approach allows us to evaluate the contribution of the different orders of scattering.

This method is particularly well suited to media of low density ($\tau < 2$), such as molecules and aerosols. However, it proves much less practical for highly-scattering and slightly-absorbing media, such as clouds, for which it is preferable to use methods such as the discrete ordinate or adding-doubling method.

4.5. List of radiative transfer codes

Table A.3 in the Appendix gives a non-exhaustive list of radiative transfer codes and their main characteristics and applications. These codes may be monochromatic or enable us to simulate the radiative quantities over a given spectral interval. Depending on which method is used to solve the RTE, they enable us to solve radiative transfer for 1D or 3D atmospheres. The most accurate approach is to couple a Monte Carlo-type method (to solve the RTE in 3D) with a line-by-line (LBL) code for calculating the gas absorption, enabling us to calculate the radiative quantities at a very high spectral resolution. Libraries providing data and resources pertaining to radiative transfer codes in the Earth's atmosphere (freely available under the GNU General Public License) are also referenced. An extensive list of codes, references, comparisons and resources on radiative transfer codes suited for 3D atmospheres is also accessible on the Website of the International Intercomparison of 3D Radiation Codes (I3RC)[2]. A similar exercise has been performed for 1D codes at long and short wavelengths[3].

These exercises in intercomparison of radiative transfer codes were put in place in the 1990s (ICRCCM: InterComparison of Radiation Codes in Climate Models [ELL 91]), driven by the Joint Scientific Committee of the World Climate Research Program (WCRP) and the International Radiation Commission (IRC) of the International Association of Meteorology and Atmospheric Physics (IAMAP). The purpose of these programs are evaluating and contributing to the improvement of the computations used in the climate models in the domains of short and long wavelengths. Moreover, a validation of those codes is also possible on the basis of radiative measurements. For example, we can cite atmospheric radiation measurement (ARM[4]), established by the US Department of Energy, which gives us access to a very wide range of radiation measurements, taken on the ground or in the atmosphere. These intercomparisons have shown that while high-spectral-resolution codes generally exhibit good consistency, greater differences may occur between narrow- or wide-band models. For instance, intercomparison exercises for numerous atmospheric situations have shown differences of less than 1.5 $W.m^{-2}$ on the fluxes calculated at high spectral resolution with a clear sky, integrated over the thermal infrared window.

2 http://i3rc.gsfc.nasa.gov.

3 http://circ.gsfc.nasa.gov/index.html.

4 http://www.arm.gov/.

4.6. For further information

Scattering

[BOH 07] BOHREN C.F., HUFFMAN D.R., *Absorption and Scattering of Light by Small Particles*, John Wiley and Sons, 2007.

[BOR 99] BORN M., WOLF E., *Principles of Optics*, 7th edition, Cambridge University Press, 1999.

[VAN 81] VAN DE HULST, H.C., *Light Scattering by Small Particles*, Dover Publications, 1981.

[MIS 14] MISHCHENKO M.I., *Electromagnetic Scattering by Particles and Particle Groups: An Introduction*, Cambridge University Press, 2014.

[KOK 08] KOKHANOVSKY A.A., *Aerosol Optics*, Springer 2008.

Theory of radiative transfer in the atmosphere

[CHA 60] CHANDRASEKHAR S., *Radiative Transfer*, Dover Publications Inc., New York, 1960.

[LEN 85] LENOBLE J., *Radiative Transfer in Scattering and Absorbing Atmospheres: Standard Computational Procedures*, A. Deepak Publishing, Hampton, VA, 1985.

[LEN 93] LENOBLE J., *Atmospheric Radiative Transfer*, A. Deepak Publishing, Hampton, VA, p. 532, 1993.

[GOO 95] GOODY R.M., YUNG Y.L., *Atmospheric Radiation: Theoretical Basis*, Oxford University Press, second edition, 1995.

[LIO 02] LIOU K.N., *An Introduction to Atmospheric Radiation*, Academic Press, Second Edition, 2002.

[WEN 12] WENDISCH M., YANG P., *Theory of Atmospheric Radiative Transfer*, Publisher: Wiley-VCH; first edition, 2012.

[THO 02] THOMAS G.E., STAMNES K., *Radiative Transfer in the Atmosphere and Ocean*, Cambridge University Press, 2002.

[ZDU 07] ZDUNKOWSKI W., TRAUTMANN T., BOTT A., *Radiation in the Atmosphere: A Course in Theoretical Meteorology*, Cambridge University Press, 2007.

3D atmospheric radiative transfer

[MAR 05] MARSHAK A., DAVIS A.B. (eds), *3D Radiative Transfer in Cloudy Atmosphere*, Springer, New York, 2005.

Radiative transfer in water

[MOB 94] MOBLEY C.D., *Light and Water: Radiative Transfer in Natural Waters*, Academic Press, May 27, 1994.

5

Methods of Geophysical Parameter Retrieval

Remote-sensing measurements obtained using radiometers or infrared spectrometers correspond to the electromagnetic radiation that has passed through the Earth's atmosphere. Hence, they do not give a direct measurement of the gas or particle concentration in the atmosphere, but rather of their influence on the radiation. As explained in the previous chapter, modeling the radiative transfer between the radiation source and the instrument enables us to calculate the spectral radiance observed when we know the instrument's characteristics and the state of the atmosphere. In this chapter, we look at how to solve the inverse problem – i.e. to deduce the gaseous and/or particulate composition of the atmosphere on the basis of observed data.

After defining the concept of a state vector, in relation to the forward model, we begin by looking at the simple case of the linear model, which enables us to introduce the necessary formalism that will be used in the case of nonlinear models. In particular, we will focus on the optimal estimation method, which has the advantage of easily giving access for sensitivity studies, information content and error budget analysis; in this sense it is one of the most effective methods. Finally, we shall examine alternative inversion methods, which employ lookup tables (LUT). These methods, whilst they are less elaborate, exhibit the advantage of being easy to employ, and requiring shorter computation times. Hence, these methods are most commonly used in the case of atmospheric inversions.

5.1. Inversion process

There are numerous approaches for retrieving geophysical parameters, commonly called inversion processes, depending on the type of measurements and the complexity of the problem. Inversion in the context of spatial remote sensing is a typical example of a problem which is mathematically ill-conditioned, meaning that it contains more unknowns than available equations, which leads to the existence of a multitude of viable solutions. This being the case, it is necessary to restrict the range of possibilities by way of *a priori* knowledge of the result. These so-called variational (VAR) methods are designed to yield the best possible estimation of the state by adjusting an initial value by means of a probability distribution. In this chapter, we will only look at the "simple" case of inversion of measurements provided by a single instrument, where we wish to determine, for instance, the vertical profile of the concentration of one or more gaseous or particulate species (1D-VAR). Although it is not discussed in this chapter, note that in the case of inversion including several instruments or the assimilation of data for the purpose of weather forecasting, it is helpful to extend the use of probability distributions in space (X, Y, Z) (3D-VAR) and time (t) (4D-VAR). Finally, it is important to note that the dependence of the result on the model and on the *a priori* knowledge, which is inherent to the inverse problem, necessitates the step of comparison/validation with other measurements – direct if possible – such as probes for water vapor and ozone, or cascade impactors, coupled with Raman micro-spectroscopy for aerosols, for example.

5.1.1. *Principle of the inversion process*

The general principle of an inversion method is to minimize the difference between an observation and the result of a model which simulates that observation. In the specific case of retrieving geophysical parameters by remote sensing, the role of the observation will be taken by the radiance spectrum corresponding to the interaction between the radiation and the atmosphere contained in the instrument's IFOV, while the model will be a radiative transfer code (see Chapters 3 and 4).

5.1.2. *The measurement vector and state vector*

The measurement vector, written as y, comprises m values, which, in the case of infrared spectra with high spectral resolution, correspond to the m radiance values measured in the spectral range covered by the instrument. The length of the vector y, i.e. the number m of radiances which will be used for the inversion, depends, first, on the spectral resolution of the instrument, and second, on the width of the spectral window of interest. That window corresponds, more or less, to the spectral range which presents the most intense absorption band for the species in question.

The state vector, written as x, is a vector with n elements that represents the set of atmospheric parameters influencing the model, which we wish to determine. These elements might include, for example, the concentration (total column or divided along the vertical) of one or more gaseous species; but also on the altitude, the thickness of the layer and the particle size distribution in an aerosol plume.

5.1.3. *The forward model*

The forward model $F(x)$ illustrates how the measurement vector y depends on the state vector x. With the inversion of atmospheric compounds, the forward model is a radiative transfer code, described in Chapter 3 for gases, and Chapter 4 for particles. It represents the best formalism in physics which links those two values in reality. This formalism involves the use of n_b parameters b of the model, which are not perfectly known and which we do not wish to (or are unable to) retrieve. Such is the case, typically, with spectroscopic parameters, or with meteorological and thermodynamic data. Thus, we have:

$$y = F(x, b) + \varepsilon \qquad [5.1]$$

where ε corresponds to the noise vector or the measuring error vector.

Thus, the inverse model I describes how the inverted parameter $\hat{x}$ is obtained from the measurement:

$$\hat{x} = I(y, b) = I[F(x, b) + \varepsilon, b] \qquad [5.2]$$

The method used to obtain the best estimation of the parameter $\hat{x}$ will be determined on the basis of the dependency of x on the forward model F. In all cases, the aim will be to minimize a mathematical function called a "cost function", which is representative of the deviation between the measurement vector y and its value as simulated by the model, $F(x) = \hat{y}$, with the possible addition of a regularization term to restrict the extent of the variability, in more complex cases – e.g. in the case of multiple solutions.

5.2. Linear models

In general, the use of a linear model corresponds to the ideal case where all of the observations y are able to be used to obtain a single, exact solution for x. In practice, for remote sensing?, this is never the case. Indeed, by its very nature, the remote sensing measurement, regardless of its geometry of observation, contains only a limited amount of information. This limitation is due in part to the technical aspects of the measurement, and to the noise with which it is marred by. It is also, however, due to the fact that certain components of the surveyed atmospheric field do not contribute to the measured spectrum (e.g. atmospheric layers where the sought component is present in only a very low concentration), yet, on the other hand, that the measurement depends on variables other than that which we are seeking to evaluate (e.g. the overlap of absorption lines of different atmospheric components, known as "interfering variables"). Hence, we cannot invert an atmospheric profile with infinite vertical resolution by using a finite number of remote sensing measurements, but only a limited amount of independent information. In mathematical terms, we say that the problem of inversion is ill-posed and under-constrained. There are more unknowns than there are equations to find them, and there is no unique solution. Nonetheless, linear models exhibit the advantage of being simple and effective for statistical studies, or for finding semi-quantitative geophysical parameters including a large amount of data.

5.2.1. *Linear least squares (LLS) method*

If we look at the simple case where the forward model F shows linear dependence on x and ε represents the residual – i.e. the difference between the measurement vector y and its simulation $\hat{y}$, then the inversion process consists of determining the value of the parameter x which best reproduces the measurement – i.e. the value of x for which the difference ε is smallest. In order to do so, we minimize the following cost function:

$$J(\hat{x}) = \sum_{j=1}^{n} \sum_{i=1}^{m} [y_i - F(x_j, b)]^2 \qquad [5.3]$$

$J(x)$can be considered to be a measurement of the distance between the m experimental data and the model which reproduces those data.

If we have an estimation of the standard deviation of the noise affecting the measurements y_i, and that noise obeys a normal distribution (Gaussian), then we can weight the contribution of the measurement to χ^2. This is a statistical test which enables us to quantitatively verify the match between the experimental values and those produced by a model. Thus, the smaller the degree of uncertainty in a measurement, the more heavily it will be weighted:

$$\chi^2 = \sum_{j=1}^{n} \sum_{i=1}^{m} \mathbf{S}_\varepsilon^{-1} \, (y_i - F(x_j, b))^2 \qquad [5.4]$$

where $\mathbf{S}_\varepsilon^{-1}$ is the inverse of the variance of noise afflicting the measurements, such that: $\mathbf{S}_\varepsilon = E[(y- \bar{y})(y - \bar{y})^T]$. This term is commonly known as the weight of the measurement, and is written as **W** in matrix form.

The values of the sought parameters x_j are then those which minimize the function χ^2, which is more usually written in matrix form:

$$\chi^2 = \mathbf{E}^T \mathbf{W} \mathbf{E} \qquad [5.5]$$

where $\mathbf{E}^T$ represents the transpose of the matrix $\mathbf{E}=\mathbf{Y}-\mathbf{F}(x)=\mathbf{Y}-\hat{\mathbf{Y}}$, with $\mathbf{Y}$ and $\hat{\mathbf{Y}}$ respectively, being the matrices corresponding to the measured values y_i and the simulated observations $\hat{y}$. It should be noted that **W** is generally a diagonal matrix, but may exhibit non-diagonal elements when there are correlations between the spectral measurements.

In order for the value of the function χ^2 to be minimal, its derivative in relation to the sought parameters must be 0, so that:

$$\frac{\partial}{\partial x_j}(\mathbf{E}^T\mathbf{W}\mathbf{E}) = 0 \qquad [5.6]$$

This, which is known as the normal equation, has the following solution:

$$\hat{X}=(\mathbf{K}^T\mathbf{W}\mathbf{K})^{-1}\mathbf{K}^T\mathbf{W}\mathbf{Y} \qquad [5.7]$$

where $\mathbf{K}$ $(m \times n) = \frac{\partial y_i}{\partial x_j}$ is the Jacobian matrix of the weighting functions and represents the sensitivity of the measurement y_i to the parameters x_j.

5.2.2. *Regularized linear model*

The regularization of an inverse problem corresponds to the idea that the observation y, on its own, is incapable of yielding an acceptable solution for x. Thus, various very different solutions for x, correct or otherwise, may be compatible with the observations. In this case, equation [5.6] has no unique solution, and the matrix ($\mathbf{K}^T\mathbf{W}\mathbf{K}$) is singular and therefore cannot be inverted. In order to circumvent this problem, it is possible to introduce a constraint on the solution in the form of data that are independent of the measurement, drawn from *a priori* knowledge of the probable solution space. This dataset, called the regularization of x, can be estimated on the basis of physical considerations pertaining to its very nature, or to some of its properties such as size or concentration. The regularization might, for example, take the form of a constraint on the variability of x or on the vertical resolution of the inverted profile. In the latter case, the inversion can be restricted to the determination of the concentration of x for the whole of the atmospheric column sounded (i.e. the total column) for the measurements at the nadir, or to a limited number of partial columns (atmospheric layers of variable thickness, for which the concentration of the component is estimated).

Mathematically speaking, this constraint is obtained by minimizing a function which is the sum of two terms: $F_1[F(x),y]$, which measures the difference between the estimation and the measured value, and $F_2(\hat{x})$, which is the penalty function including the constraint, such that:

$$J(\hat{x})= F_1[F(x), y]+ \lambda F_2(\hat{x}) \quad [5.8]$$

The Lagrange multiplier λ determines the trade-off between respecting the measurements and a constraint satisfying the solution. That value results from a compromise between the requirement of fidelity to the observations and the need for regularity postulated by the state vector.

The function $F_2(\hat{x})$ may take different forms, such as the variance $\sum(\hat{x}_i - \bar{x})^2$, with $\bar{x}$ representing the average value of x. The function $F_1[F(x), y]$ is then equal to the expression of $J(\hat{x})$ from equation [5.3].

It is also possible to replacethe penalty function $F_2(\hat{x})$ with a matrix product in the form $\hat{x}^{\mathrm{T}}\mathbf{R}\hat{x}$, where $\mathbf{R}$ is a matrix determined by the shape of the regularization function (Twomey–Tikhonov regularization).

Hence, equation [5.8] becomes:

$$J(\hat{x})= (\hat{y}-y)^{\mathrm{T}}\mathbf{S}_{\varepsilon}^{-1} (\hat{y}-y) + \lambda (\hat{x}^{\mathrm{T}}\mathbf{R}\hat{x}) \quad [5.9]$$

The minimization of J in terms of $\hat{x}$ gives us:

$$\hat{x}= \mathbf{R}^{-1}\mathbf{K}^{\mathrm{T}}(\mathbf{K}\mathbf{R}^{-1}\mathbf{K}^{\mathrm{T}} + \lambda\mathbf{S}_{\varepsilon})^{-1}y \quad [5.10]$$

Moreover, if any information x_a about the *a priori* state of the system is available, then the regularization function, which represents a measurement of the difference from the *a priori* state, is written as:

$J(\hat{x})= (\hat{y}-y)^T\mathbf{S}_\varepsilon^{-1} (\hat{y}-y)+ \lambda (\hat{x}-x_a)^T\mathbf{R} (\hat{x}-x_a)$, whose minimization in terms of x becomes:

$$\hat{x}= x_a+ \mathbf{R}^{-1}\mathbf{K}^T(\mathbf{K}\mathbf{R}^{-1}\mathbf{K}^T + \lambda\mathbf{S}_\varepsilon)^{-1} (y-y_a), \text{ where } y_a = F(x_a).$$

5.3. Nonlinear inversion

In the case of inversions of geophysical parameters on the basis of remote-sensing measurements, the vast majority of models (radiative transfer codes) used to reconstruct the measurement are nonlinear, meaning that there is no simple matrix relation such as $\mathbf{Y} = \mathbf{F}(x)$.

The existence of components of the profile which do not contribute to the measurement, and instrumental noise, are responsible for the multitude of solutions to the inverse problem. Even with low degrees of absorption, the convolution of the spectrum by the instrumental function introduces nonlinearities between the state vector x and the measurement vector y. In a moderately-nonlinear case such as this, there is no general expression for the optimal solution, and it needs to be numerically calculated by an iterative process.

Just like in the linear case, the minimum of the sum of the squares of the residuals is reached when the gradient of the cost function becomes 0. Because the problem is formulated with n parameters, we therefore have n normal equations of the type:

$$\frac{\partial J(\hat{x})}{\partial x_j} = 0 \text{ , } j \in [1, n] \qquad [5.11]$$

Nonetheless, in a nonlinear system, the differentials $\partial J(\hat{x})/\partial x_j$ depend as much on the explicative variables x as on the parameters b of the model. Hence, it is not possible to solve the normal equations as easily as in the case of a linear system. Therefore, we use a numerical solution, with a stepwise procedure of the type: $x^{k+1} = x^k + \Delta x$, which gives us successive approximations that gradually grow nearer to the true value of x.

In cases which are not too greatly nonlinear, the initial model is linearized, at each iteration, by a Taylor series around x_a, such that:

$$\mathbf{Y} = \mathrm{F}(x_a) + \mathbf{K}(x_a)\,\Delta x + \varepsilon \qquad [5.12]$$

The Jacobian matrix **K** depends on the data and the approximation adopted, so it changes from one iteration to the next.

On the basis of equation [5.4] and [5.12], we can rewrite the normal equations [5.11], whose solutions will be of the form:

$$\Delta x = \left(\mathbf{K}^{\mathrm{T}}\mathbf{W}\mathbf{K}\right)^{-1}\mathbf{K}^{\mathrm{T}}\mathbf{W}\mathbf{E} \qquad [5.13]$$

The increment Δx gives the new values of x for the next iteration, if they improve the χ^2.

In summary, nonlinear least squares (NLS) inversion is an iterative process, which requires a starting point and stop criteria, whereas the LLS are direct and depend solely in the linear algebra. Hence, NLS requires us to calculate the Jacobian matrix (first derivatives) at each step, but it may be complicated to obtain an analytical expression. In that case, a numerical differentiation, known as the finite element method (FEM), must be used. In addition, divergence is a typical problem with the NLS method. Indeed, it is not uncommon to see an increase in the cost function from one iteration to the next. This may be due to the lack of precision of the linear approximation by the Taylor series expansion. Finally, unlike with LLS, for which the solution is unique, here there may be several local minima which can hinder convergence or may converge for an incorrect value of the parameter in question.

The Levenberg-Marquardt (LM) method is a variant of the NLS method. Its purpose is not to find a linear form of the relation between y and x, but it supposes that the function χ^2 of a nonlinear model can be approximated by a quadratic equation. In this case, the solutions to the normal equations are written as:

$$\Delta x = \left(\mathbf{K}^{\mathrm{T}}\mathbf{W}\mathbf{K} + \mathbf{L}^{\mathrm{T}}\mathbf{W}\mathbf{E}\right)^{-1}\mathbf{K}^{\mathrm{T}}\mathbf{W}\mathbf{E} \qquad [5.14]$$

where $L_{i,jk} = \frac{\partial^2 \hat{y}_i}{\partial x_j \partial x_k}$, j,k=1,n and i=1,m.

Note that if we discount the second derivative terms, which is justified when the model is well suited and therefore these terms are near to the random measurement error – i.e. the second derivatives tend to assume a value of 0 when we add together the measured points – then equation [5.14] is equivalent to equation [5.13].

The Levenberg-Marquardt method is used, in particular, for the inversion of limb viewing or solar occultation measurements. Such is the case for trace gases in low concentration or which are short-lived, for which there is no *a priori* knowledge, but in this case the number of parameters is limited by the geometry of sight, which concentrates on a single atmospheric layer.

5.4. Optimal estimation method (OEM)

When the models (radiative transfer codes) used to reproduce the measurement are nonlinear and the observations y on their own are unable to provide an acceptable solution for x, as in the case of the inversion of a vertical profile of gaseous species concentrations using a spatial instrument observing at nadir, it is helpful to use an iterative, regularized method. Here, the regularization represents the contribution of the *a priori* state, which consists of the best statistical knowledge that we have on x before the measurement. The *a priori* state can be determined on the basis of climatologies or profiles measured by other instruments and enable us to restrict the components of the profile which do not contribute to the measurement. The profiles obtained are then no longer simple determinations of the true state of the atmosphere, but are instead the best estimation of it when we know the measurement and the *a priori* state of the atmosphere.

The optimal estimation method is the most widely used for the inversion of all types of geophysical parameters on the basis of measurements obtained in nadir viewing. Indeed, the measurement of an integrated column involves numerous unknown parameters along the vertical axis, which we need to regularize as best we can, using *a priori* knowledge (state vector and variability) in order to decrease the multiplicity of solutions. Moreover, this method is capable of perfectly characterizing inversions. The OEM formalism [ROD 00] enables us, first, to quantify the sensitivity of the measurement and its informational content and to separate the components of the parameters which are contained in the measurement from those drawn from the *a priori* state. Moreover, it enables us to calculate the errors resulting from the noise in the measurement and the smoothing of the real profile by the observation system (measuring instrument and inversion algorithm).

As previously stated, the inversion of remote measurements is not a direct way of obtaining information about geophysical values, then it requires a methodical approach which essentially consists of three steps:

1) first studying the sensitivity of the measurement enables us to estimate which patterns of variations of the parameters sought lead to changes in the measured values greater than the noise;

2) analyzing the informational content enables us to separate the components of the state vector which are actually derived from the measurement from those components derived from the *a priori* state. It also enables us to quantify the number of independent informational elements contained in the measurement;

3) error analysis enables us to quantify the uncertainties resulting from the inversion.

Moreover, error analysis enables us to rationalize the choice of covariance matrices which are of fundamental importance for the characteristics of the inverted data. This choice of regularization constraints is optimized to obtain the least possible inversion error by avoiding, first, too heavy a weighting of the *a priori* state that would lead to too great a smoothing error, and second, too heavy a weighting of the measurement which would cause an increase in the measurement error.

5.4.1. *Inversion method*

The optimal estimation method (or Gauss–Newton method) is based on a statistical approach to the inverse problem, which rests on the combination of the measurement and the *a priori* information using Bayes' theorem:

$$P(x|y) = \frac{P(y|x)P(x)}{P(y)} \quad [5.15]$$

Equation [5.15] comprises the following variables:

1) $P(x|y)$ is the conditional probability of x given y. P(x|y)dx represents the probability that x will have a value between x and x+dx, knowing that y has a given value. It is that probability density which is useful for the inversion of the remote measurements;

2) $P(y|x)$ is the conditional probability density of y given x, and is written as:

$$P(y|x) = \frac{1}{(2\pi)^{1/2}|\boldsymbol{S_\varepsilon}|^{1/2}} \exp\left[-\frac{1}{2}\left(y - F(x)\right)^T \boldsymbol{S_\varepsilon^{-1}}\left(y - F(x)\right)\right] \quad [5.16]$$

3) $P(y)$ is the probability density of the measurement y before it is taken. If the variable is distributed normally, the probability is Gaussian, and is written as:

$$P(y) = \frac{1}{(2\pi)^{1/2}|\boldsymbol{S_\varepsilon}|^{1/2}} \exp\left[-\frac{1}{2}(y - \bar{y})^T \boldsymbol{S_\varepsilon^{-1}}(y - \bar{y})\right] \quad [5.17]$$

4) $P(x)$ is the probability density of x before the measurement, such that:

$$P(x) = \frac{1}{(2\pi)^{1/2}|\boldsymbol{S_a}|^{1/2}} \exp\left[-\frac{1}{2}(y - x_a)^T \boldsymbol{S_a^{-1}}(y - x_a)\right] \quad [5.18]$$

where $\mathbf{S_a} = E[(x - x_a)(x - x_a)^T]$, the covariance matrix describing the *a priori* knowledge (i.e. prior to the measurement) of the state vector.

Bayes' theorem enables us to invert the forward model and obtain *a posteriori* knowledge of the state, which takes account of the information contained in the measurement. By using Bayes' theorem (equation [5.15]) and combining equations [5.16], [5.17] and [5.18], we obtain:

$$-2\ln P(x|y) = \left(y - F(x)\right)^T \mathbf{S}_\varepsilon^{-1}\left(y - F(x)\right) + (x - x_a)^T \mathbf{S}_\mathbf{a}^{-1}(x - x_a) + C \quad [5.19]$$

where C is a constant.

Equation [5.19]enables us to define the following cost function:

$$J = \left(y - F(x)\right)^T \mathbf{S}_\varepsilon^{-1}\left(y - F(x)\right) + (x - x_a)^T \mathbf{S}_\mathbf{a}^{-1}(x - x_a) \quad [5.20]$$

To find the most probable state $\hat{x}$ (maximum *a posteriori* (MAP) solution), we need to minimize J, which is again tantamount to zeroing its derivative:

$$\nabla J = \mathbf{K}^\mathbf{T} \mathbf{S}_\varepsilon^{-1}\left(y - F(x)\right) + \mathbf{S}_\mathbf{a}^{-1}(x - x_a) = 0 \quad [5.21]$$

We can attempt to find the value of x which zeros ∇J by means of the iteration equation, which, when the problem is moderately nonlinear, is written as:

$$x_{i+1} = x_a + \left(\mathbf{K}_\mathrm{i}^\mathrm{T} \mathbf{S}_\varepsilon^{-1} \mathbf{K}_\mathrm{i} + \mathbf{S}_\mathbf{a}^{-1}\right)^{-1} \mathbf{K}_\mathrm{i}^\mathrm{T} \mathbf{S}_\varepsilon^{-1}(y - F(x_i) + \mathbf{K}_\mathrm{i}(x_i - x_a)) \quad [5.22]$$

The reconstructed state $\hat{x}$ is obtained after convergence, meaning when the difference between each element of the model F between two successive iterations $|F(\hat{x}_{i+1}) - F(\hat{x}_i)|$ is less than or equal to the noise on the measurement.

Note that the contribution of the *a priori* constraints x_a and $\mathbf{S}_\mathbf{a}$ may be significant, where the measurement contains little or no information about the atmospheric profile – e.g. regarding its vertical structure or under optically thick tropospheric clouds. Furthermore, in certain cases, there is no *a priori* knowledge that is sufficiently precise, such as for the highly-reactive species given off during biomass fires or volcanic eruptions, for which there is no sufficiently-representative climatological data. In this case, the Tikhonov–Philips method is a variant of the OEM and corresponds to an iterative version of the Twomey–Thikhonov method. Here, the *a priori* covariance matrix is replaced by a regularization matrix **R** to constrain the solution.

The whole of the mathematical formalism is identical to the OEM, replacing the *a priori* covariance matrix $\mathbf{S_a}$ with the regularization matrix **R**, which enables us to carry out the same tests of the sensitivity, informational content and error budget. This method is used for the analysis of spectra obtained both with limb and nadir viewing [KEI 08].

5.4.2. *Sensitivity of the measurement and informational content analysis*

A simple way of testing the sensitivity of the measurement to the parameter in question is to calculate the Jacobian **K** and compare it to the noise on the measurement. In practice, we determine $\frac{\partial \hat{y}}{\partial x}$ analytically, in case of absorption solely, and we calculate numerically using the finite element method (i.e. $\frac{\Delta \hat{y}}{\Delta x}$) for the most complex models involving multiple scattering, for example. Then, the sensitivity is proportional to the absolute value of **K**.

However, numerous spectral channels (or absorption lines) show identical values of **K** and a spectral distribution along the vertical. This means that these values provide equivalent, redundant information. Thus, in order to more accurately determine the number of informational elements of the state vector present in the measurement, it is useful to calculate the number of degrees of freedom for signal (DoFS). Based on the above equations, we can define a matrix of the averaging functions **A** (n x n), such that:

$$\mathbf{A} = \frac{\partial \hat{x}}{\partial x} = \mathbf{GK} \qquad [5.23]$$

where **G** ($n \times m$) is the gain matrix wherein each row corresponds to the derivative of the state vector yielded in relation to the spectral channels, such that:

$$\mathbf{G} = \partial\hat{\mathbf{x}}/\partial\mathbf{y} = (\mathbf{K}^T\mathbf{S}_\varepsilon\mathbf{K} + \mathbf{S}_\mathbf{a}^{-1})^{-1}\mathbf{K}^T\mathbf{S}_\varepsilon^{-1} \qquad [5.24]$$

where $\mathbf{S}_\varepsilon = \mathbf{S}_m + \mathbf{K}_b\mathbf{S}_b\mathbf{K}_b^T$, with $\mathbf{S}_\mathbf{m}$ and $\mathbf{S}_\mathbf{b}$ being the covariance matrices, respectively, for the measuring error and the error pertaining to the parameters of the model which are not inverted.

The elements A(i,j) of the matrix **A** represent the model's sensitivity to the parameters we are seeking to determine. Each element in the inverted profile is the result of the product of the real profile by the corresponding averaging function. For an ideal observation system, **A** would be the identity matrix.

These functions generally take the form of a peak in limb viewing and are broader with a line of sight from the ground or at the nadir. The width at half-maximum of the peaks of the weighting functions (the rows in **A**, also called "kernels") gives us information about the vertical resolution of the measurement, and about the sensitivity of the spectrum in relation to each inverted parameter (sensitivity as a function of the altitude, for example). The number of informational elements contained in the measurement can then be estimated on the basis of the DoFS that are given by the plot of the matrix **A** (d=Tr(**A**)).

The generalized use of averaging kernels enables us to establish the sensitivity of the inversion for each parameter sought, and therefore make an optimal selection of those parameters by giving priority to those which have most sensitivity across the spectrum and thus are most exploitable by the radiative transfer model. Thus, we gain in terms of computation time, by limiting the number of parameters to be determined, such as the number of concentrations along the vertical, but also in terms of the quality of adjustment of the spectra and retrievals. By having fewer parameters, we avoid too great a correlation between them, which usually causes compensation effects that can lead to non-physical or erroneous solutions.

5.4.3. *Error analysis for the retrieved profile*

Inversion is characterized by explicitly evaluating the relation between the inverted profile, the true profile and the various sources of error. Therefore, we must include the measuring error in the transfer equation and take account of the fact that the inverse model uses the best estimation $\widehat{b}$ that we have for b, whereas the measuring system uses the exact values of b. Thus, the best estimation of x is written as:

$$\hat{x} = I(F(x, b) + \varepsilon, \hat{b}) \qquad [5.25]$$

After linearization and rearrangement of the terms, we deduce that the total error on the reconstructed parameter can be characterized by the covariance matrix of the total error $\mathbf{S_x}$, which is simply written as:

$$\mathbf{S}_{\mathrm{x}} = \mathbf{S}_{\mathrm{smoothing}} + \mathbf{S}_{\mathrm{meas}} + \mathbf{S}_{\mathrm{fwd.mod.}} \qquad [5.26]$$

Thus, we can see that the error is decomposed into three components:

1) The covariance of the smoothing error represents the vertical sensitivity of the measurement to the inverted profile, and is written as:

$$\mathbf{S}_{\mathrm{smoothing}} = (\mathbf{A} - \mathbf{I})\mathbf{S}_a(\mathbf{A} - \mathbf{I})^T \qquad [5.27]$$

2) The covariance of the error due to the measurement is such that:

$$\mathbf{S}_{\mathrm{meas.}} = \mathbf{G}\mathbf{S}_m\mathbf{G}^T \qquad [5.28]$$

where $\mathbf{S_m}$ is determined on the basis of the instrument's performances (radiometric bias and noise). This error matrix is commonly assumed to be diagonal, and the ith diagonal element is defined as follows:

$$\mathbf{S}_{m,ii} = \sigma_{m,i}^2 \text{ with } \sigma_{m,i} = \frac{y_i}{SNR} \qquad [5.29]$$

where $\sigma_{m,i}$ is the standard deviation of the ith spectral channel of the measurement vector y, representing the noise in the radiance; SNR is the signal-to-noise ratio.

3) The error on the non-inverted parameters of the model is such that:

$$\mathbf{S}_{\text{fwd.mod.}} = \mathbf{GK}_b\mathbf{S}_b(\mathbf{GK}_b)^T = \mathbf{GS}_\text{f}\mathbf{G}^T \qquad [5.30]$$

These errors may be random or systematic, such as, the errors on the spectroscopic data or the errors on the concentrations of the interfering species – i.e. those species whose absorption lines disturb or overlap those of the species being examined.

5.4.4. *Example of water vapor profile retrieval from IASI*

To illustrate our point, in this section we examine the example of the application of the OEM to the retrieval of a water vapor vertical profile from the IASI sounder (see Table A.1 in the appendix). In this case, the observation vector y corresponds to a radiance spectrum comprising $i = 1$ to 8600 spectral channels.

The first step is to perform a sensitivity study using the Jacobian **K** (Figure 5.1). It is calculated for $H_2{}^{16}O$ on the basis of a standard US profile and a constant surface emissivity across the whole of the spectrum y – i.e. between 650 and 2760 cm^{-1}. Here, the state vector x is composed of $j = 45$ vertical atmospheric layers of the Volume Mixing Ratio (VMR) of $H_2{}^{16}O$ between 0 and 45 km of altitude per 1 km step. From Figure 5.1, we can see that the value of **K** remains 0 beyond 2000 cm^{-1}, because that spectral range presents no absorption line for $H_2{}^{16}O$. Between 650 and 1200 cm^{-1}, the lines are less numerous, but may be interesting in the case of an inversion procedure, because they are less disturbed by absorption continuum and by the presence of absorption lines of interfering molecules [HER 09]. Thus, from Figure 5.1, we deduce that the measurement from the IASI instrument is primarily sensitive to water vapor between 1200 and 1800 cm^{-1} between 0.5 and 20 km of altitude, which defines the limits of the spectral domain and the range of altitude which will be used for the inversion.

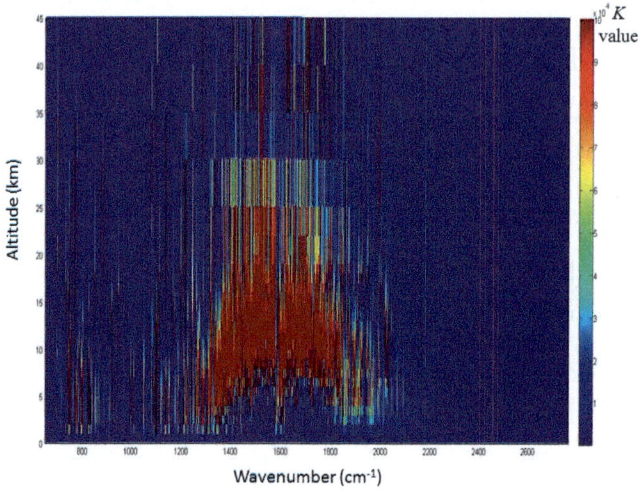

Figure 5.1. *Representation of the Jacobians* **K** *for $H_2{}^{16}O$ between 650 and 2760 cm^{-1} for a standard US profile, and the instrumental characteristics of the IASI sounder (see color section)*

The *a priori* vertical profile $\boldsymbol{x_a}$ and the corresponding variance-covariance matrix $\mathbf{S}_a$ for $H_2{}^{16}O$ (Figure 5.2) are constructed using data on the pressure, temperature and relative humidity, analyzed by the European Centre for Medium Range Weather Forecasts (ECMWF), on a vertical grid between 0 and 20 km of altitude (see Figure 5.2). This climatology covers several years, and is averaged for each season and latitudinal band: [(+90°,+60°), (+60°,+23°), (+23°,-23°), (-23°, -60°) and (-60°,-90°)]. The covariance matrix of the radiometric noise σ_ε is, in this case, diagonal, and is set on the basis of the estimated level of instrumental noise, near to the value of the root mean square (RMS) after adjustment, such that $\sigma_\varepsilon = 2 \times 10^{-9} [\frac{\mathrm{W}}{(\mathrm{cm}^2 \ \mathrm{sr} \ \mathrm{cm}^{-1})}]$.

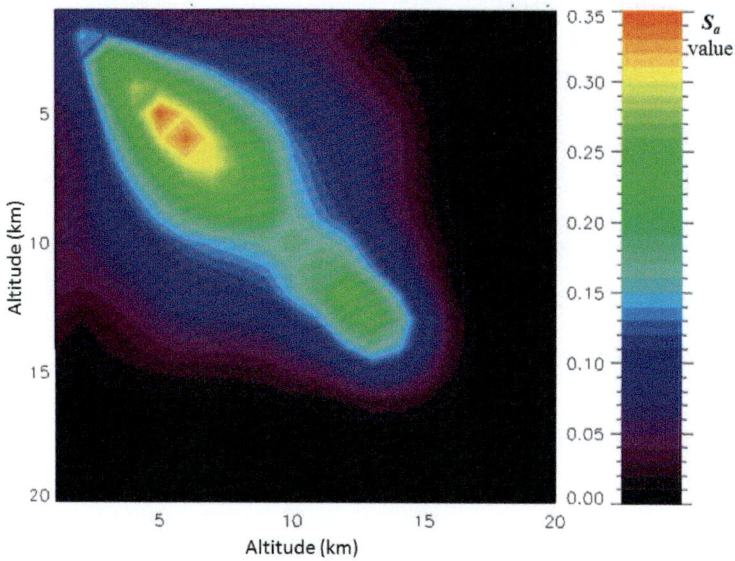

Figure 5.2. *Representation of the normalized covariance matrix* $\mathbf{S_a}$ *as a function of the altitude (see color section)*

The VMRs of the different isotopologues are given for 10 vertical layers such as: 0-1, 1-3, 3-5, 5-7, 7-9, 9-11, 11-13, 13-15, 15-17 and 17–20 km. The profiles of the interfering species (x_b) HDO, $H_2{}^{18}O$, N_2O, HNO_3, CH_4, $H_2{}^{17}O$ and CO_2 are set to their standard values in the algorithm. The surface temperature and the profiles of $H_2{}^{16}O$ are adjusted simultaneously, so they are integrated into the state vector x. In these conditions, the corresponding averaging kernels are illustrated in Figure 5.3. This figure shows that with an optimal exploitation of the measurement, we can obtain up to six levels of information along the vertical between 0.5 and 18 km. Moreover, the distribution of the weighting functions suggests that the inversion could be done for five atmospheric layers between 0 and 10 km, each of 2 km thickness, and a sixth layer between 10 and 16 km. Such a division of the atmospheric column would ensure that the inversion yields the best possible information about the partial column of water vapor between the surface and 16 km of altitude.

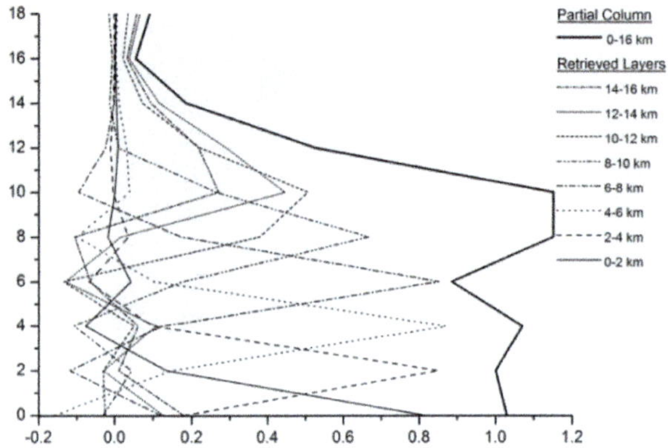

Figure 5.3. *Example of vertical distribution (0 to 16 km) of averaging kernels for $H_2{}^{16}O$ from IASI*

At this stage, we have seen that it is possible to invert a vertical profile of water vapor between 0.5 and 16 km of altitude using all of the spectral channels ranging between 1200 and 1800 cm^{-1}.

Figure 5.4(a) shows an example of the contributions of the different errors on the inversion of $H_2{}^{16}O$. The figure illustrates the error along the vertical stemming from the non-inverted parameters of the model and the instrumental noise, and the smoothing error. It is noteworthy that the measuring error is predominant up to 8 km, which is in accordance with what is observed on the averaging kernels (Figure 5.2), where the values drop sharply at altitudes of more than 8 km. On the other hand, beyond 8km, the smoothing error increases, and represents almost all of the uncertainty at an altitude of around 16 km. This means that the measurement no longer provides any information, and the value of the VMR yielded is identical to its *a priori* value. This also corresponds to what we see in Figure 5.2, where the kernel smoother is near to zero at 16 km. Figure 5.4(b) is a close-up representation of the errors linked to the interfering molecules. We can see from this figure that if the effect of the species CH_4 and N_2O is negligible, then that of the isotopologue HDO is most significant, and passes the value of 1.5% around 4 km. Such an error budget suggests that to optimize the inversion of $H_2{}^{16}O$ (i.e. to minimize the error in the retrieved concentration), it would be prudent to include the HDO profile as part of the inversion process. However, other parameters of the model will be included in the inversion, first, in order to deliver the best possible compromise between computation time and quality of the *a priori* knowledge, and second, to decrease the total error on the retrieval.

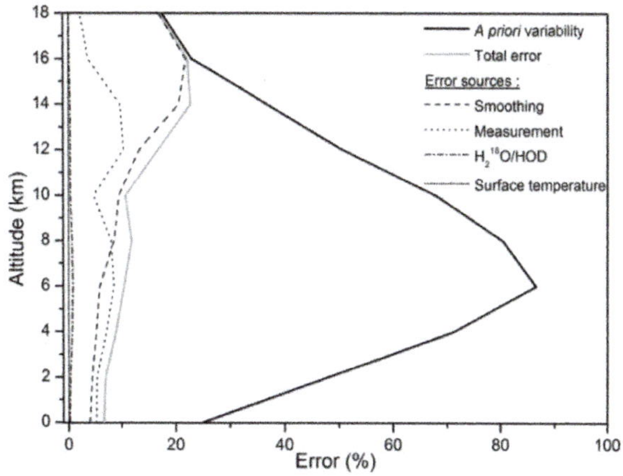

Figure 5.4(a). *Vertical profile of errors (in %) for the inversion of $H_2^{16}O$ from IASI measurements*

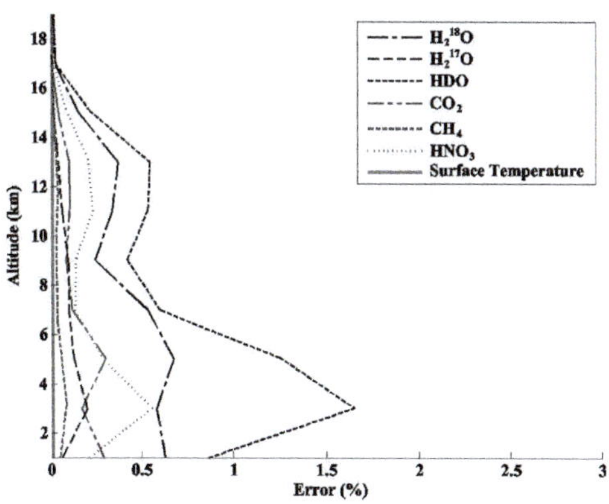

Figure 5.4(b). *Contributions of the interfering molecules in the total error (in %) along the vertical for the inversion of $H_2^{16}O$ from IASI measurements*

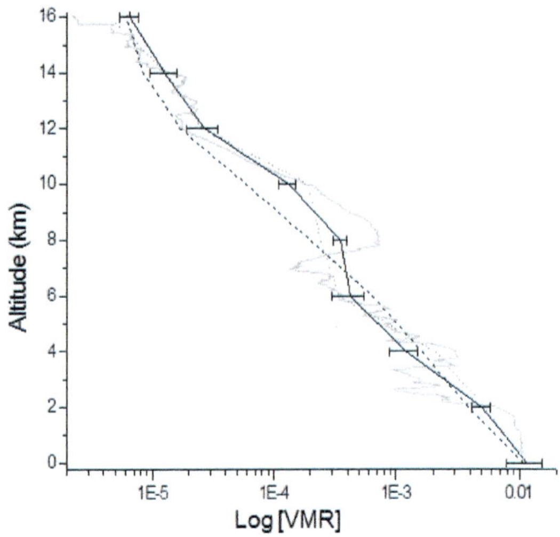

Figure 5.5. *Comparison between a retrieved profile (solid black line) with the optimal estimation method on the basis of a spectrum from IASI, and the values found by a colocalized radiosonde (solid gray line). The dotted curve represents the a priori profile x_a. The horizontal bars correspond to the uncertainties of each inverted atmospheric layer, the value of which is given by the total error $\mathbf{S_x}$*

The vertical profile thus inverted can then be compared with the values found by a colocalized radiosonde; an example is shown in Figure 5.5. It is noteworthy that in spite of a different vertical resolution and imperfect temporal (± 2 hours) and spatial co-localization (± 3° long/lat) between the two types of measurement, there is a very close correspondence between the two profiles. This result illustrates the ability of the OEM to make the best possible use of the measurements taken by infrared remote sensing at high spectral resolution in nadir viewing for the determination of the vertical distribution of the gaseous species concentration.

5.5. Lookup tables

The methods described above are generally highly effective for finding atmospheric parameters. However, they have the disadvantage of being costly in terms of computation time – particularly in the case of inversion of cloud- or aerosol parameters, which require multiple scattering calculations, as well as a good *a priori* knowledge of the system under observation. An alternative is to use a large number of pre-calculated spectra, grouped together in the form of learning tables called lookup tables (LUT's). The different calculated spectra must therefore provide as accurate a representation as possible of the variability of the parameter(s) we are seeking to estimate, but also other elements which have a significant influence, such as the surface properties (temperature and emissivity), the angle of sight or indeed the gaseous or particulate interfering species. The observed spectra y are then compared to those contained in the LUT using, for instance, the square of the Euclidean distance, which can be weighted by the covariance matrix of the instrumental noise. The sought parameter is then found by minimizing the χ^2 using equation [5.5]. There are many advantages of the LUT-based approach to retrieval: the assurance of numerical convergence, the solution obtained always has a physical meaning, and the tables can be used very quickly, meaning we can use them to process a large number of data. These advantages, and its simplicity, make it one of the most widely used methods for atmospheric sounding. Thus, the LUT-based restitution has been used on desert dust [PEY 10], clouds [YUE 09], and for the detection of trace gases such as SO_2 [PRA 07] or NH_3 [VAN 14].

For the exploitation of infrared spectro-radiometers, one of the most widely-used LUT-based methods is the "split-window" technique [INO 85, PRA 89]. This method enables us to exploit the spectral variations of the optical properties of the particles in the infrared window, and it requires us to have at least two available satellite channels in that spectral domain. It is commonly used for the filtering and retrievals of parameters for clouds [INO 87, PAR 91] or volcanic ash plumes [PRA 12, WEN 94] – situations where the use of an optimal-estimation inversion method proves very costly in terms of computation time, and is unable to assure convergence toward a physical solution, for lack of *a priori* knowledge of the state vector.

The "split-window" technique is therefore at the heart of numerous operational algorithms for processing satellite data – e.g. the imaging infrared radiometer (IIR) for studying cirrus clouds [GAR 12], whose results are available from the datacenters of NASA/LaRC (https://eosweb.larc.nasa.gov/) and ICARE (http://www. icare.univ-lille1.fr/). It is also used for the detection of volcanic ash clouds [ZEH 10], using instruments such as MODIS, AVHRR and SEVIRI. The theoretical principle of the "split-window" technique is detailed below for remote sensing of cirrus clouds. Examples of its application are presented in Chapter 6, for the characterization of volcanic ash plumes and ice clouds.

The "split-window" method is particularly well suited for the study of ice crystals by infrared observations, because the majority of these particles exhibit an absorption peak in the atmospheric window of thermal infrared between 8 and 14 µm, which is more or less pronounced depending on the form of the crystals (Figure 5.6). From this spectral signature, we can derive a relation between the brightness temperature T_B measured in a channel and the brightness temperature difference (BTD) in two channels. For example, Figure 5.7 illustrates the variation of the BTD between the channels at 10.6 and 12 µm as a function of the brightness temperature T_B at 12 µm, supposing different effective diameters D_{eff} for the crystals. Each arch corresponds to an optical thickness τ_c of the cloud varying from 0 to 50, for a given diameter of crystal. Thus, in Figure 5.7, we observe a system of arches corresponding to the relation between the BTD and T_B, each point corresponds to a given optical thickness and diameter of the crystals. Based on a couple with measured brightness temperatures, it is therefore theoretically possible to deduce from this the optical thickness of the cloud and the diameter of the crystals making it up, by interpolation of the arches. In this case, the reconstruction will be possible (or of better quality) when the concavity of the arch is most pronounced, which is generally the case for small effective diameters ($D_{eff} < 80$µm). Conversely, large and small optical thicknesses (near to the points OC and CA) and/or larger effective diameters exhibit small variations in BTD, indicating low sensitivity to τ_c or D_{eff}. The three sets of arches in Figure 5.7 correspond respectively to simulations of T_B for an observation of a cloud from space (TOA – Top of Atmosphere) or under the cloud from an airborne sensor at 3 km of altitude or from the ground (BOA – Bottom of Atmosphere). The arches corresponding to the TOA or 3 km exhibit great sensitivity; on the other hand, because of absorption by water vapor between the ground and the cloud, the arches from the BOA exhibit slight sensitivity, which is less conducive to an accurate retrieval.

In practice, the BTD's are pre-calculated using a radiative transfer code, and then tabulated in the form of LUT's as a function of the parameters being sought (τ_c and D_{eff}) and of the atmospheric or surface parameters. Next, these parameters are directly retrieved on the basis of the measured values of T_B, by interpolation in the LUT's. Note that the use of a measuring couple (BTD, T_B) generally leads to multiple solutions. The use of two pairs of channels (8.7–12 µm; 10.6–12 µm) enables us to better restrict the chosen solution, obtained by minimizing the differences on the properties retrieved for each of the couples. In addition, Figure 5.8 shows the differences of spectral variation of the BTDs as a function of the crystals' shape, and illustrates the sensitivity of the "split-window" technique to these models of crystals.

An alternative approach to LUT methods consists of using neural networks. This statistical model is inspired by the observation of actual biological neurons, and based on the principle of learning from experience [ROS 58]. The neural network comprises a set of simple computational elements (neurons) connected to the others by a weight value (synapse). During a learning phase, the neural network learns the statistical relationship between the datasets and determines the weight to be attributed to each connection between the neurons to correctly approximate that relationship. Unlike traditional methods, the neural network technique does not require a detailed understanding of the physics of the problem, or step-by-step construction of a model that accounts for these physical phenomena.

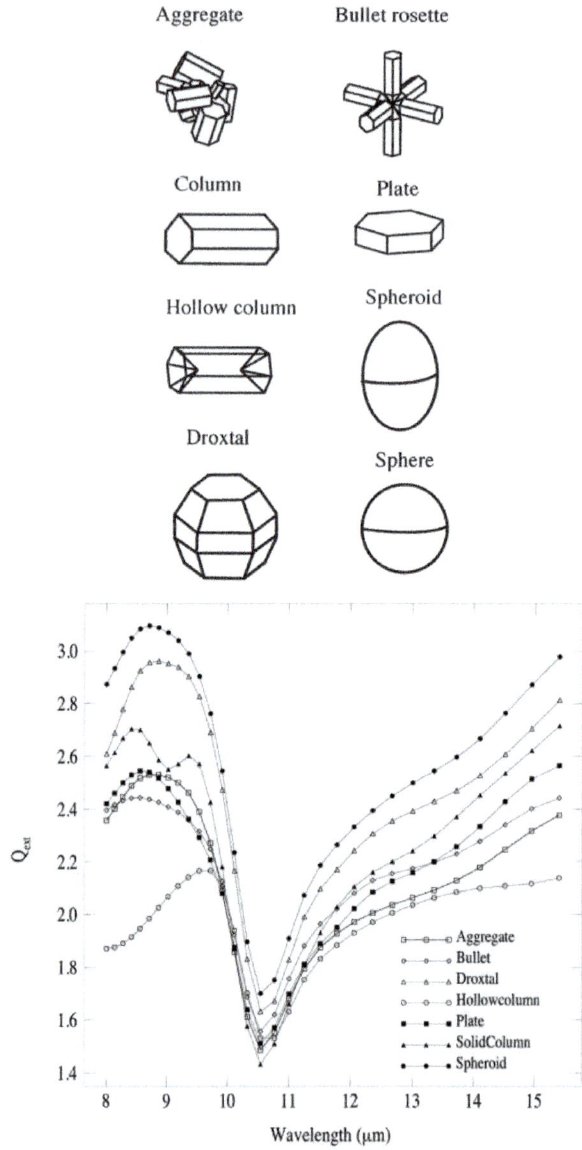

Figure 5.6. *Top: examples of crystal shapes used to model cirrus clouds (source: [YAN 05]), and bottom: spectral variation (in the infrared window) of the extinction coefficient Q_{ext} for those different crystal shapes*

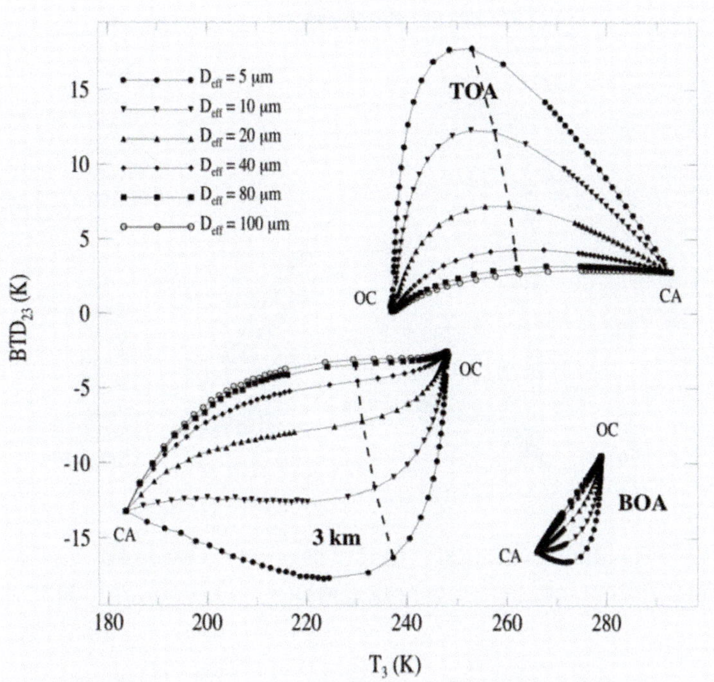

Figure 5.7. *Illustration of the "split-window" technique for a cirrus cloud situated between 9 and 10 km (237–243 K). The brightness temperature difference (BTD_{23}) between the channels at 10.6 and 12 µm is plotted as a function of the brightness temperature T_3 at 12 µm, supposing there are different effective diameters D_{eff} for crystals of fixed shape. Each arch corresponds to a given size of crystal and an optical thickness at 12 µm of the cloud varying between 0 and 50. For these arches, the points labeled CA and OC refer to the case where there are no clouds or an opaque cloud. The three sets of arches correspond respectively to simulations for an observation of the cirrus cloud from space (TOA), or under the cloud from an airborne sensor at 3 km of altitude or from the ground (BOA). Thus, each point corresponds to a given optical thickness and size of the crystals; the dotted line represents the points with an optical thickness of 2. Source [DUB 08]*

This type of neural network, in theory, can be used to approximate any function [HOR 89]. In particular, "multilayer perceptrons" enable us to solve highly nonlinear problems such as those encountered in the domain of the atmospheric sciences. They have notably been applied to the problem of inversion of atmospheric profiles on the basis of remote sensing measurements [AIR 01, MUL 03].

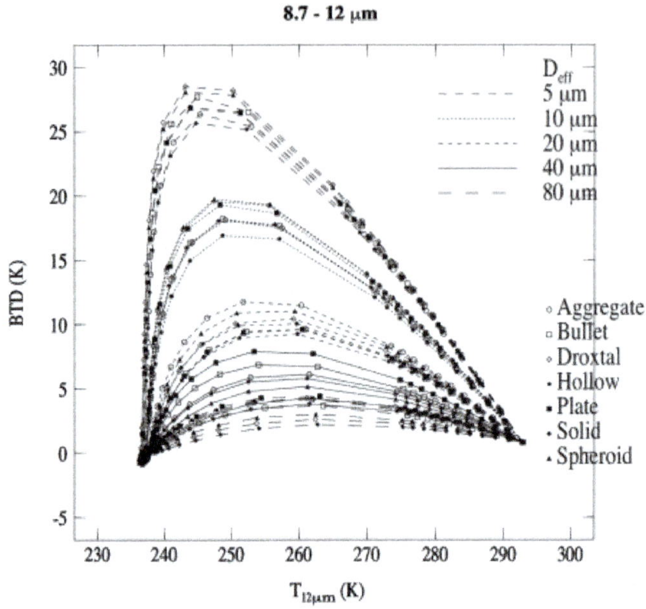

Figure 5.8. *BTD for the pair of channels [10.6, 12 μm] as a function of T_B at 12 μm, simulated for different sizes and shapes of crystals in the case of observation from the TOA*

The main advantage of neural networks is that they do not require detailed simulation of the radiative transfer leading to the measured spectra. Only a general understanding of the measuring technique is needed to determine the most appropriate parameters which, statistically, enable us to cover the greatest possible number of cases. Once their training is complete, these algorithms are also much quicker to execute than variational methods, which require complex calculations for each inversion.

However, like LUT-based methods, neural networks exhibit the disadvantage of having difficulty in characterizing the biases and errors of the parameters obtained. Furthermore, the quality of the retrievals will depend heavily on the training procedure, and therefore on the set of training profiles chosen. Given that the training set is necessarily limited, we may see results which physically make sense but are totally false – e.g. in the case of extreme weather events (inversion of the temperature profile, pollution peak, desert storms, biomass fires, volcanic eruptions, etc.).

Hence, this method is reserved, at research level, for non-quantitative detection, or more operatively, for finding stable parameters where the large amount of data can be averaged in space or time in the assimilation models.

In closing, let us note another approach, called the predictor method, which consists of exploiting the slow and uniform variation of the spectroscopic parameters with the temperature and the relation of proportionality of the optical thickness with the density of the gases. It is based on regressions for a given level of pressure. The optical thickness of absorption $\tau_{i,j}$ in a spectral channel i, defined from the top of the atmosphere at a level j, is approximated by a linear combination of variables known as predictors:

$$\tau_{i,j} = \tau_{i,j-1} + Y_j \sum_{k=1}^{K} a_{i,j,k} X_{j,k}, \qquad [5.31]$$

where $a_{i,j,k}$ are the coefficients and the K predictors ($X_{i,k}$ and Y_j), which are dependent upon the atmospheric variables (p, T, density of the gas) and the mass of air factor. The coefficients are pre-calculated on the basis of a training set of atmospheric profiles representing the atmosphere's natural variability. The set of predictors used is described in the scientific report on the RTTOV radiative transfer code [SAU 10].

5.6. For further information

[ATK 97] ATKINSON P.M., TATNALL A.R.L., "Introduction neural networks in remote sensing", *International J. Remote Sensing*, vol. 18, 1997.

[BLA 09] BLACKWELL W.J., CHEN F.W., *Neural Networks in Atmospheric Remote Sensing*, Artech House, 2009.

[ROD 00] RODGERS C.D., *Inverse Methods for Atmospheric Sounding: Theory and Practice*, World Sci., Hackensack, NJ, 2000.

[TIK 63] TIKHONOV A., "On the solution of incorrectly stated problems and a method of regularization", *Dokl. Acad. Nauk SSSR*, vol. 151, pp. 501–504, 1963.

6

Space Infrared Remote Sensing: Some Applications

Spatial remote sensing represents a considerable advance in atmospheric research. It constitutes the only means of observation from the local to the global scale, with a temporal resolution of only a few minutes for geostationary satellites. These characteristics mean that spatial remote sensing is an indispensable tool for atmospheric studies with applications in meteorology, tropospheric chemistry and air quality monitoring. It also allows for dynamic tracking of events such as biomass fires, desert dust storms or indeed volcanic eruptions, and enables us to better evaluate their possible consequences, both in terms of radiation or in terms of the environment or health. In addition, space agencies are concerned not only with improving the quality of the measurements by innovative and increasingly-effective instruments, but also with the continuity of the observations which yield access to available data spanning several years, or even decades. This allows us to study the evolution of the climatic variables such as surface temperature or concentration of greenhouse gases (CO_2, CH_4), but also to establish a climatology using numerous geophysical parameters.

Thus, after thirty years of existence, spatial remote sensing is reaching maturity, and the numerous applications covered by atmospheric research mean that it is now at the heart of societal issues.

The purpose of this chapter is not to give an exhaustive state of the art about the potential of infrared satellite observations, but to illustrate, by means of a few examples, the diversity of applications, covering the study of gases, aerosols and clouds using measurements at low and high spectral resolution.

6.1. Water vapor isotopologues

Water has the peculiarity of being present in abundance in the earth's atmosphere in all three phases (liquid, solid and gas), which, because of the water cycle, makes it a key element in the Earth's system. The water which is present in gaseous form, resulting from evaporation (from the seas and oceans) or transpiration (biomass), has a residence time in the troposphere of one to two weeks, so it can be transported over thousands of kilometers.

There are numerous approaches for measuring atmospheric water vapor. Thus, columns of H_2O can be measured from space, using visible-UV instruments such as GOME-2 or POLDER, whilst vertical profiles can be obtained using a microwave instrument (SSM/I) and Infrared (TES, IASI). One of the special characteristics of these instruments is that they are able to spectrally distinguish between different isotopologues of water vapor (see Figure 6.1). The measurements of the isotopological composition of the water vapor are used in studies of the water cycle [BEC 03], the paleoclimate [JOU 97], the physics of clouds [WEB 03], exchanges between the troposphere and stratosphere [KUA 03], or the evolution of the climate and meteorology (precipitation, storms, hurricanes) [BOW 03]. The advantage of using isotopological ratios stems from their high dependency on the conditions of evaporation and the condensation historic of the air masses during their transport. This process is known as the vapor pressure isotopologue effect (VPIE), which corresponds to the fact that heavier isotopologues have a lower saturating vapor pressure than the lighter isotopologues (e.g. $H_2{}^{16}O$), which leads to isotopologic separation during the successive phase changes. Thus, when the vapor and condensed phases are at thermodynamic equilibrium, the lighter isotopologues condense more slowly and evaporate more quickly, thus increasing the concentration of the heavy isotopologues in the liquid or solid phase. The consequence of VPIE is that the concentration of heavy isotopologues steadily decreases over the course of the transport (horizontal, but particularly vertical transport) of the water vapor, and becomes lower than the standard mean ocean water (SMOW) isotopic level.

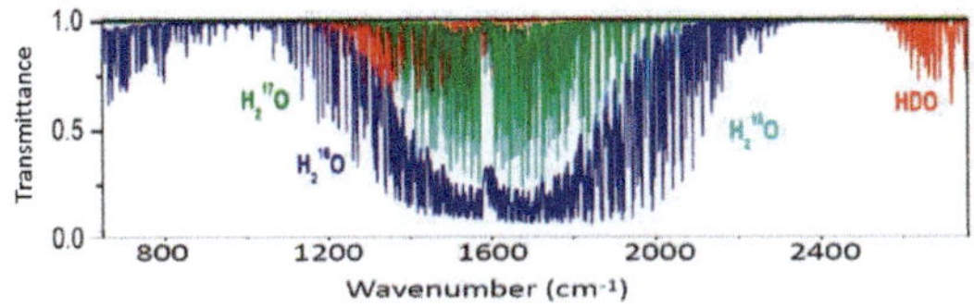

Figure 6.1. *IASI transmittance spectrum illustrating the absorption bands of the isotopologues: $H_2{}^{16}O$, $H_2{}^{18}O$, $H_2{}^{17}O$ and HDO (from [HER 07]) (see color section)*

A good knowledge of the horizontal and vertical distributions of the water vapor isotopologues is, therefore, essential in order to better understand the phenomena which govern the water cycle. However, the significant variability of their concentration, over space and time, represents a major difficulty, which explains why, to date, very few measurements of vertical profiles have been obtained. In particular, measurements taken from space are especially scarce; however, we can cite the HDO measurements in the high troposphere/lower stratosphere with the atmospheric trace molecule spectroscopy (ATMOS) instrument [MOY 96], and latitudinal distributions around 4 km of altitude (maximum sensitivity, see Chapter 5) obtained with the instrument IMG [ZAK 04]. It was with the latter that the greatest advance was made, enabling us to simultaneously obtain the horizontal and vertical distributions of the species ($H_2{}^{16}O$, $H_2{}^{18}O$ and HDO) on the global scale [HER 07], illustrated by Figure 6.2. From this figure, it can be remarked upon that the highest concentrations are found in the inter-tropical belt, as illustrated by the latitudinal (vertical) distributions, which also express a significant impact of surface temperature and the altitude of the tropopause. Maps of global (horizontal) distributions show that the concentrations of $H_2{}^{16}O$ and $H_2{}^{18}O$ are higher above the continents, at the tropics, in particular for Amazonia and Central Africa. This contrast between Ocean and Continent is less marked in the case of HDO, for which we only see a latitudinal variation, stemming from the progressive decrease of its concentration during precipitations and the transport of the air masses toward the Poles.

In addition, for the different applications of isotopologic measurements of water vapor, the essential parameter is the relative isotopic ratio. It is typically measured in relation to the SMOW and is expressed in [ppm], such that:

$$\delta_x = \left(\frac{R_x}{R_{xSMOW}} - 1\right) \times 1000 \qquad [6.1]$$

where R_x is the isotopic mass ratio of the species, R_{xSMOW} is the corresponding standard isotopic ratio – for example: $R_{18OSMOW}$ = $H_2{}^{18}O/H_2{}^{16}O$= $2.0052.10^{-3}$ and R_{DSMOW} = HDO/ $H_2{}^{16}O$ = $0.31152.\ 10^{-3}$.

The main source of atmospheric water vapor is oceanic, and the vapor is characterized by a relative isotopic composition δ=0%. The separation associated with the exchange reaction at equilibrium is expressed in terms of the factor α:

$$\alpha = R_C / R_V \qquad [6.2]$$

where R_C and R_V are the isotopic ratios of the condensed and vapor phases, respectively. The order of magnitude of α depends on several parameters, but the most important is the temperature. The separation leads to typical values of $\delta^{18}O = -12‰$ and $\delta D = -85‰$ just above sea level. When the air rises and cools, the heavier isotopologues are the first to condense, which decreases the isotopic ratio and therefore leads to negative values of δ. Thus, the value R_V of the remaining vapor gradually decreases, which, in the context of the approximation of thermodynamic equilibrium, is described by the Rayleigh equation:

$$R_V = R_0 f^{(\alpha-1)} \qquad [6.3]$$

With R_0 being the initial isotopic ratio of liquid water, f the vapor fraction and α the factor of separation at equilibrium over the course of evaporation. The isotopic composition of the water vapor is thus affected by numerous complex meteorological processes which characterize its origin.

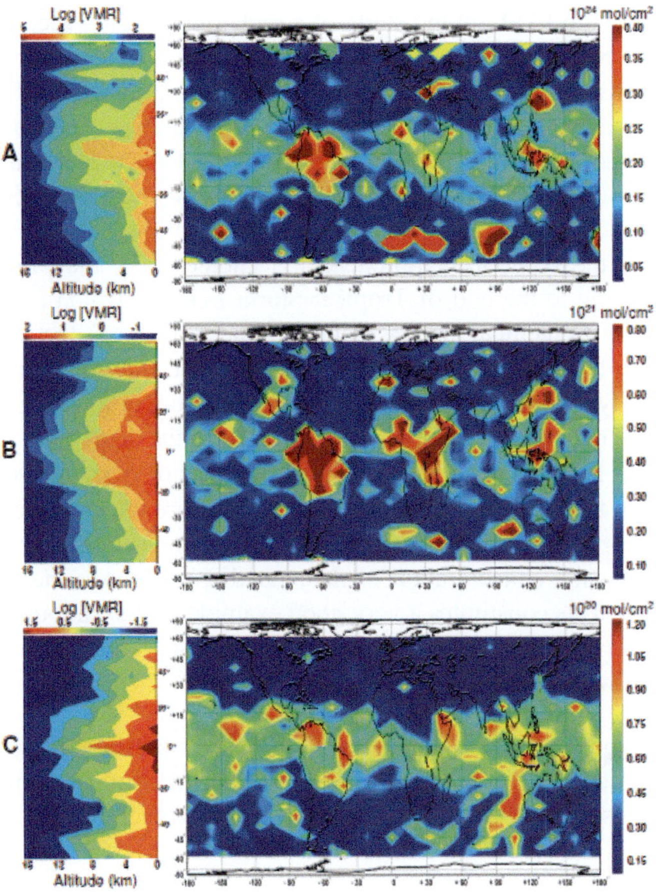

Figure 6.2. *Latitudinal distributions of the volume mixing ratio (VMR) [ppm] as a function of the altitude and latitude (left) and horizontal distributions on the global scale (right) of the concentrations [mol.cm^{-2}] of $H_2{}^{16}O$ A), $H_2{}^{18}O$ B) and HDO C) (from [HER 09]) (see color section)*

Measurements of δ_D are an excellent tracer of the structure, evolution, and water budget of cyclones, storms and hurricanes. However, previously, they were limited to the rainwater gathered on the surface or water vapor collected by an airplane. The first global distribution of the ratio δ_D (HDO/$H_2{}^{16}O$) was obtained using TES [WOR 06], whilst the first application at local scale [HER 09] was performed with IASI, on the study of a typhoon, which is accompanied by complex hydrological processes. This study yielded the spatial distributions of $H_2{}^{16}O$ and δ_D (Figure 6.3), corresponding to Typhoon Krosa, which was one of the major meteorological events of 2007, near the Philippines, between Taiwan and Southwest China. The map of the integrated water vapor columns [mm] along the vertical between 0 and 8 km on 3 October shows two humid fronts: one to the south-west, caused by the typhoon's approach, and another over the Philippines corresponding to the end of Tropical Storm Lekima, which took place between 27 September and 5 October. The map on 8 October also shows two fronts, but this time, to the north-west, the front near to the eye of the typhoon is much more intense. It is also worth noting that the amount of precipitable water is very high near to the depression (around 150 mm). The maps representing the distributions of δ_D show evolutions opposite to those of $H_2{}^{16}O$. Thus, the regions where the values of δ_D are low correspond to high humidity values, demonstrating the sever reduction in the concentration of deuterated species once the typhoon has passed. This constitutes a very good example of the "amount effect" [HER 09], whereby the isotopic ratios are anti-correlated with the amount of water.

This type of study opens the way to very interesting prospects in terms of the use of high-spectral-resolution infrared measurements to study atmospheric events – particularly major weather events.

6.2. Biomass fires and trace gases

Vegetation fires can be caused by human activities (intentionally, criminally, or accidentally), or indeed by natural factors (volcanic eruptions or lightning, for example). A lack of precipitation, or a deficit of humidity, favor forest fires, which represent a major source of aerosol- and gas emissions into the atmosphere [AND 01]. Those gases are highly diverse, and notably include CO, greenhouse gases (CO_2, CH_4, N_2O), but also numerous hydrocarbons and aromatic compounds (benzene (C_6H_6), toluene (C_7H_8), furan (C_4H_4O)), volatile organic compounds (HCOOH, CH_3OH, CH_3COOH and HCHO), along with nitride (NO_x, NH_3, HONO and HCN), sulfide (SO_2, H_2S) and halide compounds. For the most part, these emissions are confined to the boundary layer but, during ageing, the plumes may be transported both horizontally and vertically, and thus reach the upper troposphere [HOB 03]. Hence, fires have a major influence, locally, on the atmospheric chemistry, air quality and, consequently, human health. However, in certain cases, the species formed may be directly injected into the upper troposphere- lower stratosphere, in intense pyro-convection processes [FRO 06]. Therefore, fires are involved in numerous environmental problems, including weather forecasting.

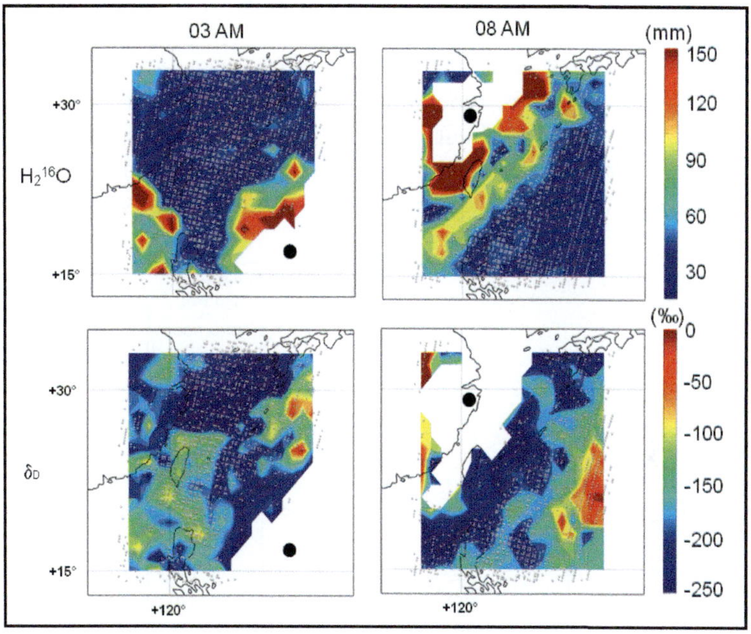

Figure 6.3. *Horizontal spatial distributions of $H_2{}^{16}O$ (integrated column along the vertical between 0 and 8 km of altitude in mm of precipitable water) and δD [‰] for the mornings of 3rd and 8th October 2007.The black disk represents the eye of the typhoon. The white parts correspond to the cloudy areas. The gray dots represent the IASIIFOVs. The maps are averaged over a latitude/longitude grid of 1° (from [HER 09]) (see color section)*

Studies of biomass burning use numerous techniques from the ground or from airplanes [YOK 03], as well as modeling, including transport and photochemistry [MAS 06]. The use of satellite measurements is more recent, and is mainly based on radiometric analyses of thermal infrared images, using imagers such as MODIS and AVHRR, or in geostationary orbit, such as SEVIRI (MSG). Thus, it is possible to detect thermal anomalies (hot points) [GIG 99], using brightness temperatures generated on the basis of the near infrared and thermal infrared spectral bands. The algorithm therefore measures thermal emission in MODIS pixels, taking account of the variabilities of the surface temperature and the reflection of solar radiation. When that heat emission surpasses the background emission threshold, the algorithm identifies an active fire. It is also possible to obtain information about the burned surface areas [FRA 00], based on vegetation indices, such as the BAI (for *Burn Area Index*) [MAR 98]. However, although this type of measurements has already yielded important information about the composition and spatiotemporal evolution [EDW 06], such techniques are limited by the lack of vertical resolution and sensitivity to gaseous species with weak absorbance.

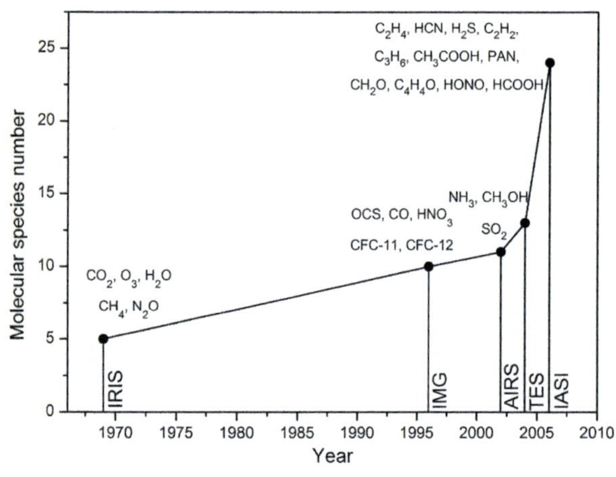

Figure 6.4. *Evolution over time of the number of trace gases measured from space by high-spectral-resolution infrared spectrometers (IRIS, IMG, AIRS, TES and IASI)*

The use of high-spectral-resolution IR spectrometers in nadir viewing constitutes a major advancement in the measurement of trace gases (Figure 6.4), including those generated by biomass. Thus, by measuring the concentrations of these different species, we are able to establish emission rates and thus determine precisely what type of fires we are dealing with. In the case of these instruments, the most commonly used method to detect the plumes from forest fires is to measure carbon monoxide (CO), which is a relatively stable atmospheric species. It is primarily produced by combustion, and a large amount of it is emitted by forest fires, which makes it a very good indicator. Thus, a biomass fire generally results in a rapid, local increase in the CO concentration. By careful analysis of the corresponding spectra, we can therefore detect and track the horizontal and vertical distributions of numerous trace gases and short-lived species.

For instance, the study of a fire on the east African coast, with the instrument ACE-fts, was the first to detect the presence of numerous compounds such as C_2H_4, H_2CO, C_3H_6O, PAN or C_3H_4 [COH 07], whose spectral signature is illustrated in Figure 6.5.

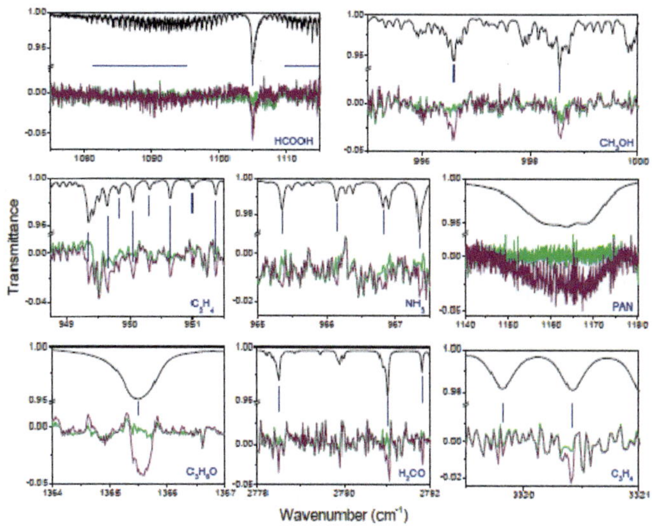

Figure 6.5. *Spectral adjustment of one ACE-fts spectrum corresponding to a tangential altitude of 11 km, illustrating the detection of weakly absorbing compounds: CH_3OH, HCOOH, NH_3, C_2H_4, H_2CO, C_3H_6O, PAN and C_3H_4. The green and purple colors represent the residual, after adjustment, respectively with and without account taken of the absorption of the species in question. The spectral contribution specific to each species is shown in black, and the vertical blue lines mark the position of the main absorption lines (see color section)*

Information on the photochemical age of a fire, for its part, can be obtained by simultaneous detection of non methane hydrocarbons (NMHC's) – particularly using the classification put forward by [MAU 98]. In the case of the plume sounded by the ACE-fts instrument (Figure 6.5), the presence of C_2H_4, which has a lifespan of between 12 h in the lower troposphere and 1.2 days at higher altitudes, allows us to classify the observed plume as being "recent", and conclude that it has not been transported for more than a day before being detected. In addition, this hypothesis is borne out by the detection of H_2CO and NH_3, which are direct products of the emissions from pyrolysis and slow combustion, and which both have lifespans of less than 2 days. It should be noted that these indirect indications about the age of the plume can be supported by the use of chemistry-transport models to estimate its origin. Such a model is a code which enables us to quantify the evolution of a plume of pollutants over time on different spatial scales (horizontal and/or vertical) on the basis of meteorology and emission-flow data.

Finally, the vertical profiles retrievals of these new species, for each spectrum between the tangential altitudes of 5 and 20 km, is represented in Figure 6.6. "Recent" plumes of smoke are generally observed near to the source in the free troposphere, as is the case here. However, all of the reconstructed profiles represented in Figure 6.6 show that the maximum value of *VMR* is located at a high altitude, around 11 km, for almost all species. This implies that the biomass fire under examination has been the object of rapid and significant uplifting. On the basis of the measured concentrations, it is also interesting to compare the relative increase in short-lived species in comparison to their background level. Thus, we find that at the altitude of 11 km, the compounds C_2H_2 and HNO_3 show significant increases: +2200% and +1175%, respectively.

6.3. Volcanic eruptions

Volcanoes are a major source of aerosols and gaseous precursors of secondary aerosols which have a significant effect on infrared and solar radiation, with consequences for the radiative forcing and the climate [LET 07]. In addition, ash clouds or acid component plumes have an impact on air traffic [PRA 09; International Civil Aviation Organization (ICAO), 2001]. In this context, satellite instrumentation represents a highly effective tool for spatiotemporal monitoring of the plume's evolution. The case of the eruption of the Icelandic volcano Eyjafjallajökull, which took place between April and May 2010, is a good example of the consequences of this type of eruption. The studies of this eruption demonstrated the advantage of remote sensing for the precise determination of the structure and dispersion of the emissions, as well as the advantage of combining those measurements with atmospheric modeling to characterize the physico-chemical properties of volcanic plumes, and predict their propagation [BOI 13].

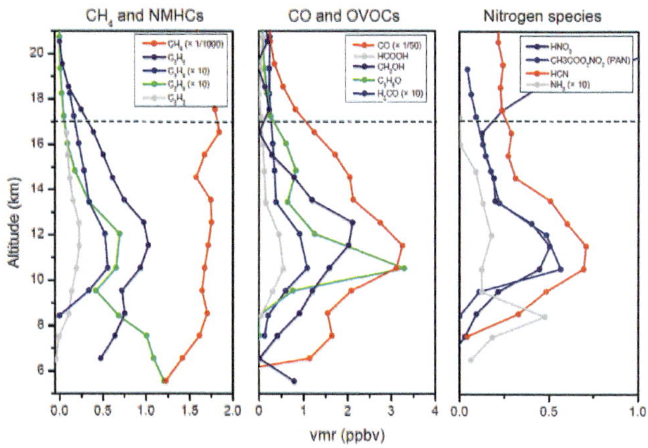

Figure 6.6. *VMR's vertical profiles (5–20 km) of the gaseous species given off by the biomass burning event. The horizontal dashed line marks the altitude of the tropopause (see color section)*

6.3.1. *Sulphur dioxide*

The atmospheric budget of SO_2 is dominated by anthropogenic missions, which represent 75% of total emissions. The main natural sources are the biogenic activity of the oceans and particularly volcanic eruptions. Thus, sulphur dioxide (SO_2) is the major gaseous component in volcanic eruptions, of which the enormous concentrations ejected into the atmosphere have a direct impact on the environment and on human health. In addition, its atmospheric lifespan makes it a good tracer of volcanic plumes over the course of several days. Generally, the absorption signal of SO_2 is weak, making it impossible to detect in a background conditions. However, although it is still delicate, it becomes possible to measure the SO_2 by satellite when there is a huge increase in its concentration, particularly at high altitude, as is generally the case with volcanic eruptions. By way of example, Table 6.1 allows us to compare the total columns of SO_2 measured by IASI during Eyjafjallajökull's eruption, with the official products of the instruments GOME-2 and OMI functioning in the UV-visible spectral range. This table recaps the results of colocalized observations (±85 km) around Tórshavn (Faroe Islands) on 13 May 2010. The differences observed in terms of the minimum concentrations are attributable to the difference in sensitivity between the instruments. For medium and high concentrations, the correspondence is very close considering the criterion of colocalization, the retrieval uncertainty and the concentrations, which are fairly weak. These results demonstrate the ability of these three instruments, with their respective processing algorithms, to retrieve the SO_2 concentration, even at modest emission rates. This local comparison also demonstrates the complementarity provided by the use of several instruments to monitor the transport of volcanic plumes over a large spatial and temporal scale.

		GOME-2	IASI	OMI
Time, UT		10:55	11:10	12:25
Number of values		12	21	15
Mean distance, km		70	50	35
SO_2 (DU)	Mean	1.76	2.66	2.53
	Min	0.14	1.80	1.42
	Max	3.38	4.26	4.17

Table 6.1. *SO_2 total columns, co-localized above Tórshavn (Faroe Islands) on 13 May, obtained using the instruments: OMI (Sulphur Dioxide Level 2G Global Binned -OMSO2G: http://disc.sci.gsfc.nasa.gov/Aura/data-holdings/OMI/omso2g_v003.shtml), GOME-2 (O3M-SAF data: http://o3msaf.fmi.fi/) and IAS*

The spatial distributions of the SO_2total columns (in Dobson units, 1 D.U. = 2.69×10^{20} mol.m^{-2}), recorded from IASI (day and night), and those found by the operational product of level 2 OMI are represented in Figure 6.7, for the days of 6 and 13 May. For each day, the overall shape of the plume is similar for both instruments. Thus, the different crossing times highlight its temporal evolution. Moreover, by combining the observations of IASI and OMI, we can obtain a better description of the transport linked to the atmospheric dynamic. It can also be remarked upon that in the case of both eruption phases, the plume, which is only tropospheric, was transported a long way from the volcano, reaching the southern Mediterranean.

Finally, note that recent studies [BOI 14] have demonstrated the capacity of coupling of remote sensing and models to find the emission flows over time.

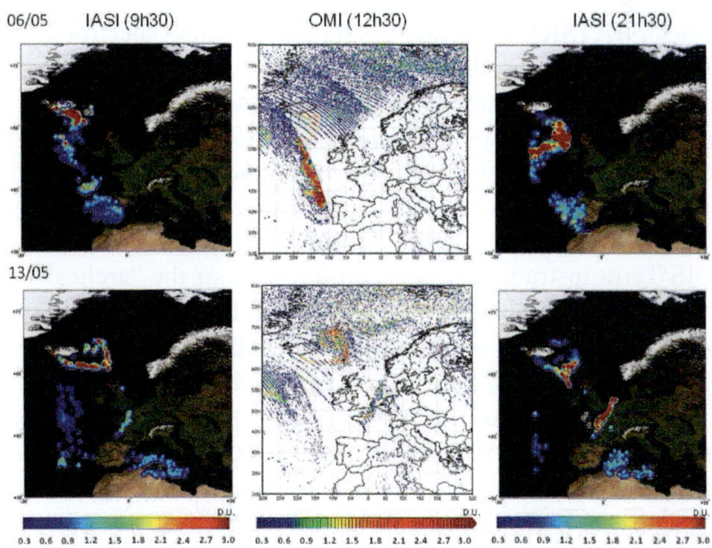

Figure 6.7. *Spatial distributions of the SO_2 total columns obtained by IASI (left and right) and OMI (middle) for 6th and 13th May. The IASI data are interpolated by a Cressman function (with R = 96 km). The OMI data are represented from the IFOV of the instrument (see color section)*

6.3.2. *Volcanic aerosols*

Volcanic eruptions also represent a major source of aerosols, whether in the form of ash, which has a high mineralogical diversity

[BON 11], or of acid droplets – particularly H_2SO_4. Volcanic aerosols, which can be ejected into the troposphere or into the stratosphere, have a significant impact on the radiative budget, but also on the environment (e.g. acid rain) or air traffic.

Numerous methods have been put forward for space-based observation of aerosols from volcanic plumes [THO 10, ZEH 10]. One of the most widely used is the so-called "split window" technique, presented in Chapter 5, and commonly used for studying volcanic particles [PRA 89]. This technique, which makes use of the sensitivity of infrared between 8 and 12 µm to the characteristics of aerosols, has already been proven effective for radiometers, such as: advanced very high resolution radiometer (AVHRR), along track scanning radiometer (ATSR), moderate resolution imaging spectroradiometer (MODIS), and for sounders in geostationary orbit, such as the geostationary operational environmental satellite (GOES) or spin enhanced visible and infrared imager (SEVIRI). Figure 6.8 illustrates the capacity of the "split window" technique to discriminate cloudy scenes from those where aerosols are present in the case of the eruption of the Eyjafjallajökull volcano, with images provided by the MODIS/Terra instruments. Thus, on the basis of the "arches" formed by the brightness temperature differences ΔBT at11 and 12µm spectral channels, it is possible to retrieve the optical thickness τ_a of the plume, and the effective radius r_e [µm] of the particles, and from those data, we can estimate the surface mass M [$g.m^{-2}$], which is an important parameter for aerial safety. Figure 6.9 shows the spatial distributions of those three parameters, retrieved from measurements of the instruments: MODIS/Terra (at 11:55 UTC), SEVIRI (at 12:00 UTC) and IASI (morning orbit) which is used here as a spectro-radiometer [DUB 14]. As the algorithm used is exactly the same and the times of observation are very close together, direct comparison of these results illustrates the differences caused by the instruments' characteristics (orbit, spatial resolution and spectral resolution). The spatial distribution of the three parameters (τ_a, r_e and M) is relatively similar for all three instruments. This is quantitatively confirmed by the mean values of the optical thicknesses, the effective radii and the surface masses obtained (Table 6.2), which are coherent considering the sources of uncertainty and the amplitudes.

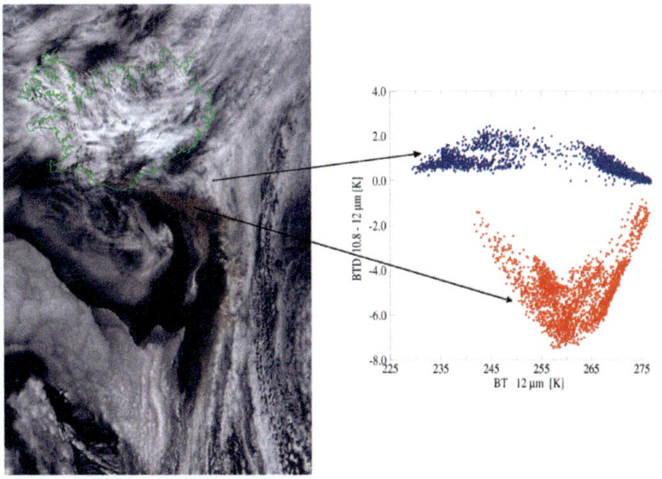

Figure 6.8. *Example of brightness temperature differences ΔBT [K] (right) during the eruption of the volcano Eyjafjallajökull, observed with the satellite MODIS/Terra (left) on 6 May 2010.The representative ΔBT of the volcanic ash plume are shown in red; those in blue correspond to to a cloudy area. ΔBT are defined as the difference between the brightness temparature at 10.8 and 12 micrometre. Source: [DUB 14] (see color section)*

Satellite	τ_a	r_e (µm)	M (g.m^{-2})	M_T (kt)
SEVIRI	0.42	4.2	1.9	162
IASI	0.54	4.4	2.3	200
MODIS	0.53	4.7	2.6	221

Table 6.2. *Mean values of the optical thickness τ_a, the effective radius r_e [µm], and the surface mass M [g.m^{-2}], obtained by the instruments (SEVIRI, MODIS and IASI) throughout the day of 6 May 2010*

The split window technique, therefore, is a simple and effective method to detect aerosols produced by volcanic eruptions. However, it is incapable of distinguishing the chemical- or mineralogical composition of the aerosol particles, whose diversity results in a significant variability of the optical properties.

A more elaborate method, but more costly in terms of computation time, is to exploit an optimal estimation method (see Chapter 5) and spectra at high spectral resolution, whose sensitivity to the type of particles has recently been demonstrated [CLA 10, HER 13]. Indeed, the spectral variability yielded by high-resolution infrared spectrometers is characteristic of the nature of the particles emitted, and therefore of the volcano from which they were erupted, as illustrated by Figure 6.10. This method of measurement, which is still in its infancy, should soon be able to give us access to the physico-chemical properties of the ejected particles and their evolution during transport, and thus better quantify their role in the atmospheric system.

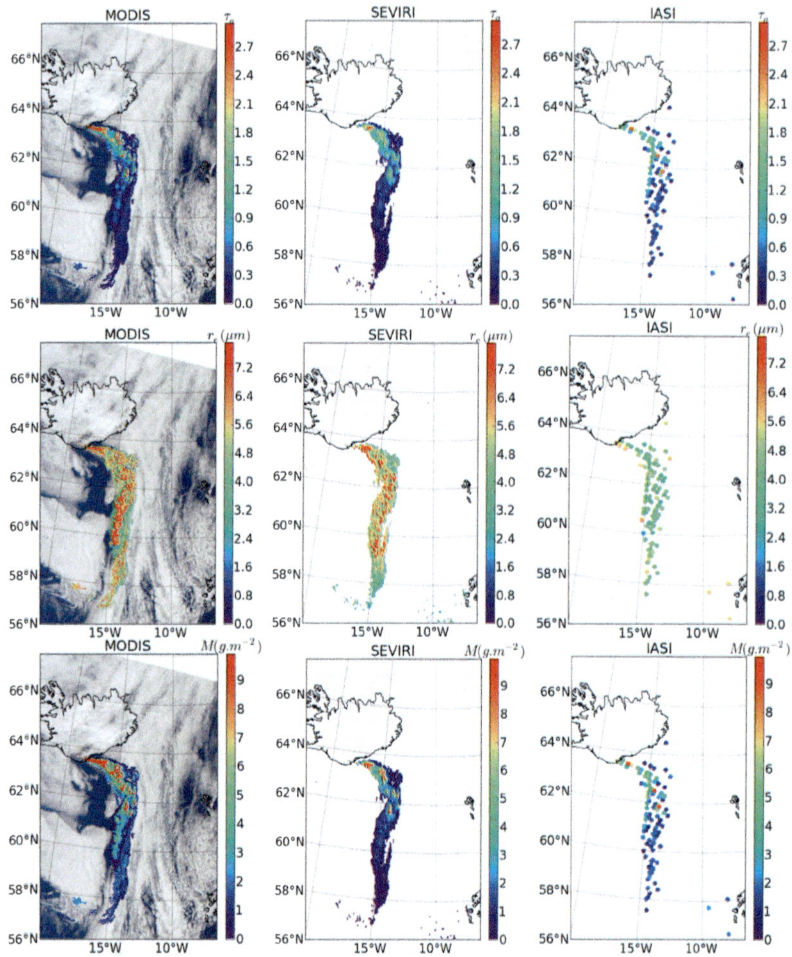

Figure 6.9. *Optical thickness τ_a at 12 μm (top row), mean radius r_e (middle row) and surface mass M (bottom row) retrieved on 6 May, using MODIS/Terra (11:55 UTC), SEVIRI (12:00 UTC) and IASI (late morning) (see color section)*

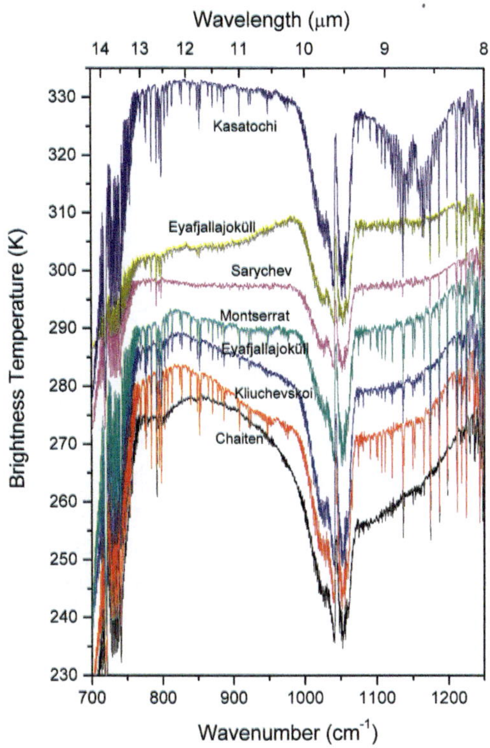

Figure 6.10. *Examples of brightness temperature spectra observed by IASI during several volcanic eruptions, illustrating the spectral variability of the plumes due to the diversity of the distributions of the particles emitted in terms of their size distribution (PSD) and type (chemical and mineralogical composition). These spectra are staggered on the ordinate (vertical) axis for ease of viewing. Adapted from [CLA 10] (see color section)*

6.4. Physical properties of clouds

Clouds play an essential role in terms of the global radiation balance in the atmosphere and, even more markedly, on weather conditions locally. Hence, it is necessary to have a precise view of their optical- and microphysical properties in order to accurately estimate their effects on the solar- or infrared radiation and, therefore, on the forecasts of meteorology or climate models. For this purpose, space observation represents a powerful tool, capable of characterizing clouds on the scale of the whole planet, with fine spatial and temporal resolution. Besides the quality of the measurements, the joint use of different types of measurements helps to improve the precision of the retrieved cloud properties. An example of such instrumental synergy is offered by the instruments of the Afternoon Train (A-Train) mission, launched in 2002. The A-Train is made up of a constellation of satellites in a circular and quasi-polar heliosynchronous orbit at an altitude of around 700km, crossing the equator at 13:30 local time, whose trajectories are spaced only a few minutes apart (Figure 6.11). In particular, the A-Train includes multiple instruments operating in the infrared, capable of – amongst other things – observing and characterizing clouds (see Table A.2a in the appendix). Also, instruments such as the cloud profiling radar (CPR) on the satellite CLOUDSAT, operating in the hyper frequency spectral domain, yield additional measurements for the study of the cycle of clouds and of water vapor. By way of example, Figure 6.12 shows a convective cloud system observed by the A-Train instruments over the African coast in June 2006 (Figure 6.12(a)). The back scattered LIDAR signal, measured by CALIOP (Figure 6.12(b)), is sensitive to small atmospheric particles. It is able to detect semi-transparent cirrus clouds around 15 km of altitude, situated at the top of the convective system and made up primarily of crystals. Also, at the edge of the cloud system, we can see a layer of aerosols situated between 3 and 5km, and some low clouds at around 3 km. However, the LIDAR beam is not powerful enough to pass through the dense clouds situated beneath the cirrus clouds, which have high content of liquid water or ice. The brightness temperatures (T_B) measured by the infrared radiometer IIR are presented for the 12 μm channel (Figure 6.12(c)). By comparison with the measurements made by CALIOP, we can verify on the IIR swath that the coolest T_B values correspond to high clouds, and the warmest, at the edge of the cloud, to aerosols and low clouds. Finally, the radar measurements of CLOUDSAT (Figure 6.12(d)) are, for their part, particularly sensitive to high content of liquid water or ice in clouds situated between 1 and 12 km of altitude, and thus enable us to compute the vertical profiles of those clouds. However, they are not affected by the semi-transparent high clouds. This example clearly illustrates the complementarity of a multi-instrument analysis for the characterization of clouds, which enables us to form a complete description of the cloud structure.

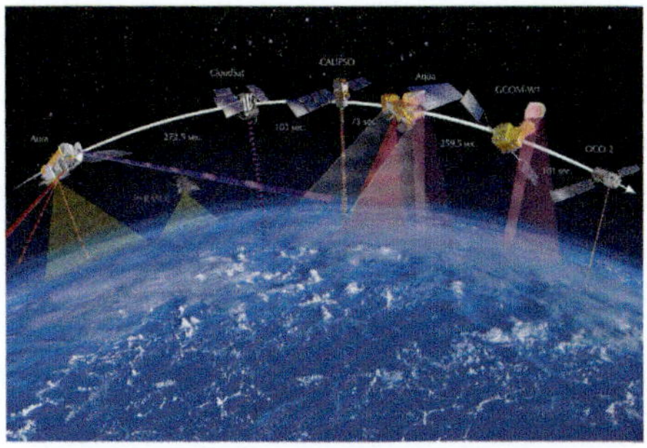

Figure 6.11. *A-train constellation of satellites in its 2013 configuration. Source: "A-Train w-Time2013 Web" by NASA JPL – http://oco.jpl.nasa.gov/ galleries/ gallery spacecraft/. Licensed under Public Domain via Wikimedia Commons – (see http://commons.wikimedia.org/wiki/File:A-Train_w-Time2013_ Web.jpg#/media/File:A-Train_w-Time2013_Web.jpg) (see color section)*

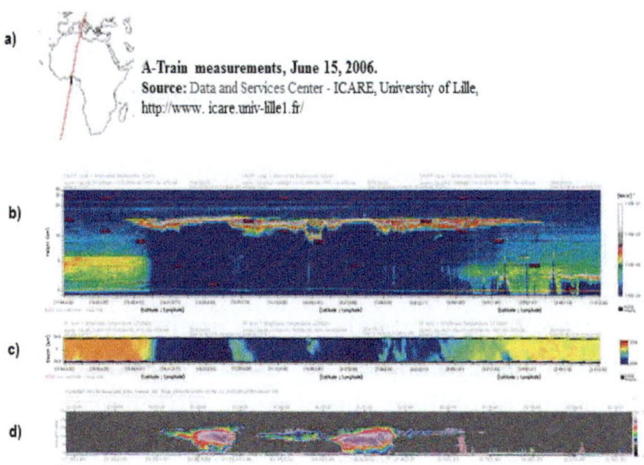

Figure 6.12. *Observation of a convective cloud system on 15 June 2006 above the Gulf of Guinea by the A-Train satellites: a) the red line represents its ground track, and the black line corresponds to the observed scene. This figure presents, as a function of the latitude/longitude; b) the vertical profile of the back scattering coefficient measured by CALIOP, showing high semi-transparent clouds around 15 km of altitude, and low aerosols and clouds around the edge of the cloud system; the red/gray zones indicate a high density of particles, the green/yellow zones a lower concentration and the dark blue zones indicate that the backscattered signal is null; c) the brightness temperature measured at 12 μm by IIR along its swath, cooler (blue) for high clouds and warmer (orange/yellow) for the aerosols and low clouds; d) the vertical profile of the measurements of CLOUDSAT/CPR, sensitive to the liquid-water and ice content (red/pink for high concentrations) of the clouds located between 1 and 12 km, below the semi-transparent cirrus clouds (see color section)*

Below, we discuss a few examples of applications of remote sensing of clouds with the A-train's infrared instruments – firstly for optical and microphysical characterization of the high-altitude ice clouds, and secondly for determination of the thermodynamic phase and the altitude of the clouds.

6.4.1. *Classification and physical properties of ice clouds*

At high altitude, the thin clouds composed mainly of ice (cirrus clouds) are fairly transparent to solar radiation but greatly absorbent of infrared. Therefore, they have a significant greenhouse effect, and the quantification of that effect depends on the optical properties, and thus microphysical properties, of the crystals which make them up. As we saw in Chapter 5, infrared remote sensing is particularly well suited for the study of crystals, because ice exhibits a pronounced absorption peak between 8 and 14 µm, in the infrared atmospheric window. The intensity of that peak depends on the microphysics of the crystals [DUB 08]. With its three channels centered at 8.7, 10.6 and 12 µm, the IIR radiometer is well configured for the study of this type of clouds. Indeed, the analysis of the brightness temperature measured in two channels situated in the infrared window enables us to estimate certain physical properties of the crystals by the method known as the split window technique (see Chapter 5). This method is based on the relation linking the brightness temperature difference (BTD) in two channels to the brightness temperature (TB) measured in a given channel. Note that, similarly, the spectral signature of cirrus clouds can also be analyzed in terms of the difference in effective emissivity ε_{eff} of the cloud. This method can easily be implemented in a data-processing algorithm. The BTD's are pre-calculated using a radiative transfer code [DUB 05], for realistic conditions that are representative of cirrus clouds, to construct pre-calculated lookup tables (LUT's) as a function of the cloud's optical thickness, as well as the size and shape of the particles that make it up. Then, these cloud parameters are directly inverted on the basis of the IIR measurements, by interpolation in the LUTs.

An operational algorithm for the measurements of the IIR sensor was developed on the basis of the split window method [GAR 12,

GAR 13], with the aid of the NASA Langley Research Center. The products are available from the NASA/LaRC (https://eosweb.larc.nasa.gov/) and ICARE (http://www. icare.univ-lille1.fr/) datacenters. The algorithm enables us to invert IIR measurements for ice clouds, at an altitude greater than 7 km – the altitude above which it is considered that the clouds are primarily composed of ice. This algorithm, firstly, enables us to establish a cloud classification for an observed scene, on the basis of the level-2 geophysical products provided by the data-processing algorithm of the LIDAR instrument CALIOP, which gives precise information about the altitude of the cirrus clouds (Figure 6.13(a)). The classification corresponding to the CALIOP profile (Figure 6.13(b)) mainly shows situations of single-layer clouds (types 21, 30, 40, and 42), referenced in the scene classification established for IIR (Table 6.3). It is then extended to the whole of the IIR swath (Figure 6.13(c)). Based on this classification, the microphysical and optical properties can be inverted. For example, Figure 6.14 illustrates, as a function of the effective emissivity at 12 μm and for the orbits during January 2011, the effective diameter D_e of the ice crystals, whose mean value ranges from 20 to 40 μm. This study helped to demonstrate the advantage to using simultaneous, colocalized infrared and LIDAR radiometric measurements to study clouds. Note, finally, that other sensors, such as MODIS, SEVIRI or AVHRR, also have similar spectral channels which are suitable for the use of the split window technique.

6.4.2. *Thermodynamic phase and altitude of clouds*

Determination of the thermodynamic phase (liquid/ice) is essential for the inversion of the optical and microphysical parameters of clouds and for predictive models. For this purpose, it is possible to make joint use of the peculiarities of the instruments POLDER and MODIS by way of three methods which have individually proved their potential for studying clouds. These methods are illustrated in Figure 6.15 for a scene corresponding to Typhoon Nabi off the coast of Japan in September 2005, and are able to yield additional information about the cloud structures [RIE 10]:

1) Use of multidirectional and polarized measurements from POLDER (Figure 6.15(a)), enabling us to tell the difference between the two phases: the polarized angular signature of clouds composed of liquid water (red) and ice (blue) exhibit a significant difference, which facilitates a simple and direct determination of the cloud's thermodynamic phase (Figure 6.15(b));

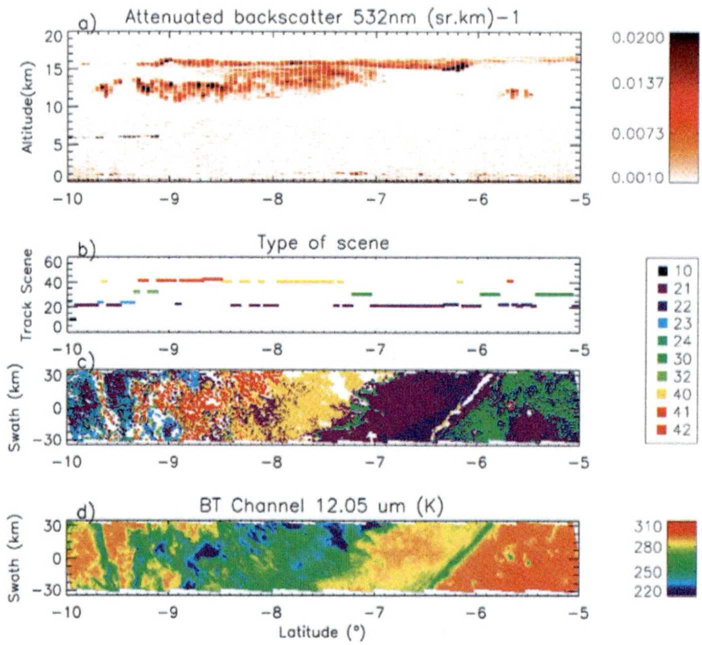

Figure 6.13. *Example of cloud classification as a function of the latitude for a scene observed by IIR: a) vertical profile of the back scattering coefficient measured by the LIDAR CALIOP; b) cloud classification corresponding to the above LIDAR profile, the color code for which is explained in Table 6.3; c) classification extended to the whole of the plot observed by IIR; d) brightness temperature at 12 μm along the length of the observed plot. Source: [GAR 12] (see color section)*

2) Use of the relation between the visible and NWIR reflectances measured by MODIS (Figure 6.15(c)): these relations can easily be configured in the form of LUTs as a function of the geometry of observation and the cloud parameters (Figure 6.16). However, it

should be noted that combining these reflectances does not always enable us to tell the two phases apart (see the shaded area in the figure);

3) Use of thermal infrared measurements from MODIS using a "split-window" method (Figure 6.15(d)): MODIS has the channels in the infrared window, so this method can be employed (see Chapter 5).

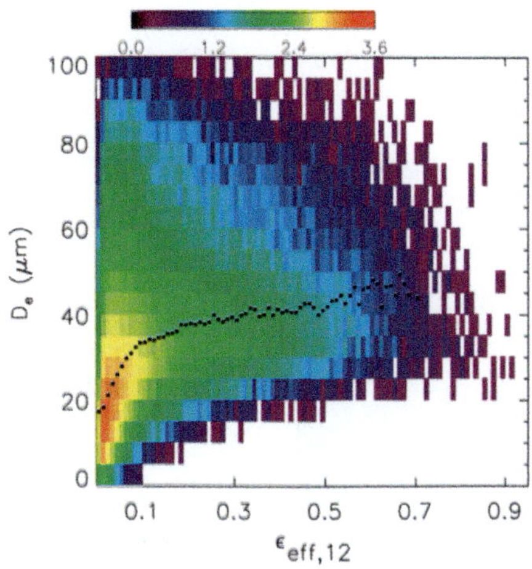

Figure 6.14. *Effective diameter D_e retrieved using the IIR measurements as a function of the effective emissivity at 12 μm $\varepsilon_{eff,12}$ for single-layer cirrus clouds, between the latitudes 60°S and 60°N during the month of January 2011. The color code represents the point density and the black dotted line shows the median value for D_e, which lies approximately between 20 and 40 μm depending on the value of $\varepsilon_{eff,12}$. Source: [GAR 13] (see color section)*

Scene	Target description	Reference
21	1 high STC layer and no aerosol layer	Surface
30	1 high STC layer and nondepolarizing aerosols	Surface
40	1 high opaque cloud layer, VDR_max > 40%	Surface
80	1 high opaque cloud layer, VDR_max < 40%	Surface
22	2 high STC layers	Surface
26	3 high STC layers	Surface

31	1 high STC layer	Low opaque cloud
32	2-5 high STC layers	Low opaque cloud
37	1 high STC layer	Low opaque aerosol
41	1 high STC layer	High opaque cloud
42	2 high STC layers	High opaque cloud

Table 6.3. *IIR scene classification for high-altitude clouds (altitude > 7 km), overlying the surface or a dense opaque layer of cloud or aerosol STC = Semi-transparent cloud and VDR = Volume Depolarization Ratio (adapted from [GAR 12])*

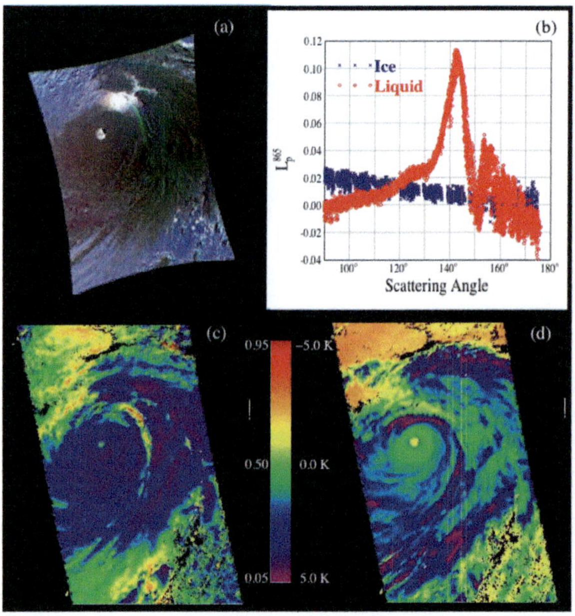

Figure 6.15. *Illustration of methods using MODIS and POLDER data to determination of the thermodynamic phase of clouds for a scene during Typhoon Nabi, on 2 September 2005 off the coast of Japan: a) scene used, artificially colorized using the polarized POLDER measurements at 490, 670 and 865 nm; b) typical signature of polarized multi-angular radiance measured by POLDER for a cloud of liquid water (blue) or ice (red) as a function of the scattering angle; c) ratio of MODIS measurements in the near-infrared (2.1 μm) and visible (0.865 μm); d) brightness temperature difference (BTD) of the MODIS channels at 8.5 and 11 μm. Source: [RIE 10] (see color section)*

These methods can easily be integrated into data-processing algorithms, in the form of LUT's , facilitating quick and functional use of the spatial measurements.

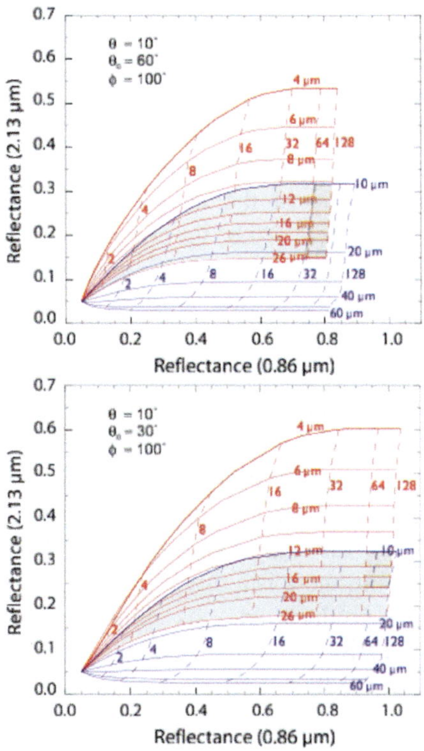

Figure 6.16. *Simulation of the reflectances at 2.13 μm as a function of those at 0.86 μm for two geometries of observation (solar zenith angle θ_0, observation angle θ and azimuth angle ϕ), for an ice cloud (blue) or liquid cloud (red), for different optical thicknesses (ranging between 0 and 128) and effective diameters of the particles (4–26 μm). The shaded area in the figure indicates that the combination of those wavelengths does not enable us to tell the phases apart. Source: [RIE 10] (see color section)*

Moreover, studies have shown that the exploitation of spatial measurements in the absorption band of dioxygen at 760 nm, called the O_2-A band, enables us to estimate the atmospheric pressure at the surface [BRE 96], the pressure at the top of the clouds [FIS 91] or the altitude of the aerosols [DUB 09]. POLDER has two spectral channels positioned around 760 nm (Figure 6.17(a)). The first channel, centered at 765 nm, has a significant spectral width (FWHM (Full Width at Half Maximum) = 40 nm), and therefore the mean absorption of O_2 over its spectral width is slight; the second, centered at 763 nm, is narrower (FWHM) ≈ 10 nm, and is therefore highly sensitive to absorption by O_2. The ratio R of the reflectances measured in these two channels, $R = \rho^{763}/\rho^{765}$, therefore approximately represents the transmission of the radiation in the O_2-A band. This transmission is a function of the length of the optical path of the radiation in the atmosphere – i.e. the air mass factor m. If we take scattering to be negligible, this factor is defined on the direct path from the Sun to the satellite as $m = 1/\cos\theta_0 + 1/\cos\theta_v$, with θ_0 and θ_v being the solar zenith angle and the satellite's angle of sight. The transmission also depends on the number of O_2 molecules found along the path, and therefore on the altitude (or pressure) at which the radiation is back scattered toward space, which generally corresponds to the altitude of the top of the cloud if it is sufficiently thick. This pressure is known as the apparent pressure of the cloud top and is written as p_a. As the dioxygen concentration in the troposphere is assumed to be constant, there is a direct relation which exists between the ratio R and the product mp_a^2, as illustrated by the simulations shown in Figure 6.17(b), computed using a high-spectral-resolution code such as a Line-by-line (LBL) code. If we know the satellite's geometry of observation, then it is possible, on the basis of the measured value of R, to determine the apparent pressure p_a (or altitude) of the top of the cloud. The use of an LBL code is recommended, because the determination of p_a is highly sensitive to R and the spectral response of both channels needs to be precisely taken into account in order to calculate the absorption of O_2 corresponding to each channel. It should also be noted that the apparent pressure p_a retrieved can differ from the real top of the cloud if it is semi-transparent or highly diffusive. Indeed, radiation can penetrate the cloud to a greater or lesser depth before it is scattered toward the satellite, thus altering the optical path and therefore the factor m. In this case, the apparent pressure p_a retrieved is higher (lower altitude) than that at the top of the cloud.

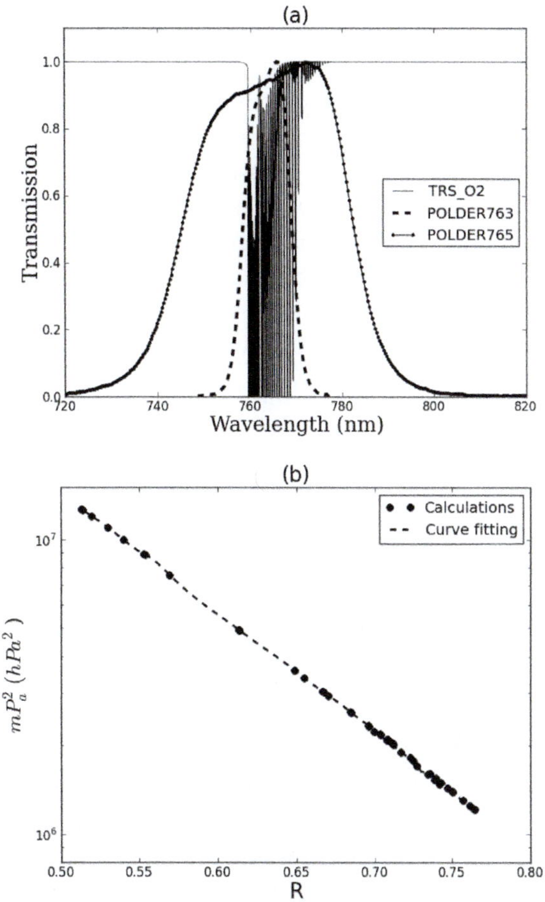

Figure 6.17. *a) Spectral transmittance of dioxygen (TRS_O2) of the wide channel (at 765nm) and narrow channel (at 763 nm) for POLDER, centered in the O2-A band around 760nm; b) mp_a^2 as a function of the reflectance ratio R= ρ^{763}/ρ^{765}, in the POLDER channels at 763 and 765 nm, where p_a is the apparent pressure of the top of the cloud and m the air mass factor. The adjustment of the curve $mp_a^2 = f(R)$ is represented by the dotted line*

Figure 6.18 shows the determination of the thermodynamic phase and the altitude of clouds, using the combination of POLDER and MODIS measurements. Note that, as expected, clouds classified as being formed of ice are present at low pressures. Most of the cloudy pixels show a temperature between 238 K and 268 K, at which the ice and supercooled-water phases can coexist. Using this methodology, it is then possible to determine this classification on the global scale. For example, Figure 6.19 shows the degree of cloud cover deduced from measurements of the POLDER sensor, the MODIS sensor or a combination of the two. The inversions obtained from POLDER and MODIS are coherent, but exhibit significant differences as to the cloud cover over certain geographic regions, due to the specifications of the instruments (spatial resolution, spectral response, contribution of polarization, limitation of the methods used, etc.). Thus, generally speaking, instrumental synergy helps to improve the inversions, by exploiting the specific strong points of each instrument. It also enables us to find a quality index as regards the inverted products, as a function of the results obtained by each instrument.

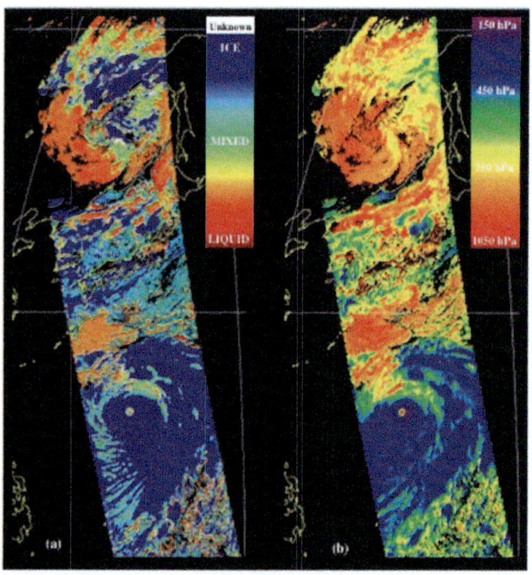

Figure 6.18. *a): Classification of the cloudy phase by combination of measurements from POLDER and MODIS; and b): determination of the cloud top pressure using measurements in the POLDER channels centered in the dioxygen band around 760 nm, for the cloudy scene on 2 September 2005 presented in Figure 6.15(a). Source: [RIE 10] (see color section)*

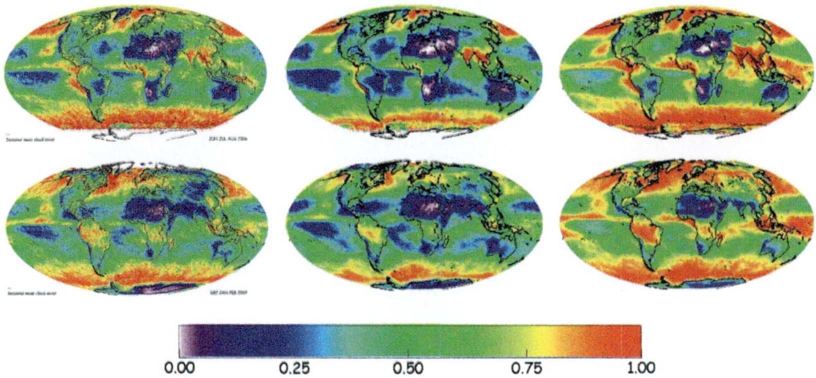

Figure 6.19. *Degree of cloud cover (from 0 to 1) deduced from measurements taken from POLDER (left-hand column), from MODIS (right-hand column) or the combination of the measurements taken by POLDER and MODIS (middle column), for the months of June, July and August 2006 (top row) and for December 2006, January and February 2007 (bottom row). Source: LOA & ICARE (see color section)*

6.5. For further information

[BOR 12] BORRELL P., MAY BORRELL P., BURROWS J.P. *et al.*, *Sounding the Troposphere from Space: a New Era for Atmospheric Chemistry*, Springer Science & Business Media, 2 February 2012.

[BUR 11] BURROWS J.P., PLATT U., BORRELL P., *The Remote Sensing of Tropospheric Composition from Space*, Springer Science & Business Media, January 15, 2011.

[CHÉ 12] CHÉDIN A., CHAHINE M.T., SCOTT N., *High Spectral Resolution Infrared Remote Sensing for Earth's Weather and Climate Studies (Nato ASI Subseries I:)*, Springer, Softcover reprint of the original 1st ed. 1993 ed., 7 January 2012.

[KHO 09] KOKHANOVSKY A.A., DE LEEUW G., *Satellite Aerosol Remote Sensing Over Land*, Springer-Verlag Berlin Heidelberg, 2009.

[KUE 13] KUENZER C., DECH S. (eds), *Thermal Infrared Remote Sensing: Sensors, Methods, Applications*, Remote Sensing and Digital Image Processing series, Springer, Netherlands, 2013.

[LEN 13] LENOBLE J., REMER L., TANRE D. (eds), *Aerosol Remote Sensing*, Springer-Verlag Berlin Heidelberg, 2013.

Appendix

Spectral Region	VHF	UHF	Microwave	Infrared	Visible	Ultraviolet	X-rays	γ-rays
Usage	NMR	EPR	Rotational transitions	Vibrational transitions	Electronic transitions		Ionisation	Nuclear effects
Wavenumber (cm^{-1})	0.017	1.0	10	1000	20000	40000	1.0×10^{7}	5.0×10^{8}
Wavelength	0.6 m	1 cm	1 mm	10 μm	500 nm	250 nm	1 nm	20 pm
Frequency (Hz)	$5.\times10^{8}$	$3.\times10^{10}$	$3.\times10^{11}$	$3.\times10^{13}$	$6.\times10^{14}$	1.2×10^{15}	$3.\times10^{17}$	1.5×10^{19}
Single photon energy (eV)	2.07×10^{-6}	1.24×10^{-4}	1.24×10^{-3}	1.24×10^{-1}	2.5	5	1.24×10^{3}	6.2×10^{4}
Photon energy ($kJ.mol^{-1}$)	2.03×10^{-4}	1.20×10^{-2}	1.20×10^{-1}	12	239	479	1.2×10^{5}	6×10^{6}
Main remote sensing applications	**High-frequency radar:** *H_2O & cloud profile, wind velocity, surface altitude*			**Spectroradiometry:** *gas spectrometry & profiling, cloud & aerosol properties*	**Photopolarimetry:** *cloud & aerosols, surface reflectance, O_3, trace gas and aerosols*			

Table A.1. *Conversion table. The wavelength λ and wavenumber $\tilde{\nu}$ are linked by the formula $\tilde{\nu} = 1/\lambda$, with the following conversion relations: $\tilde{\nu}$ [cm^{-1}] = $10^4/\lambda$ [μm] and $\Delta\tilde{\nu}/\tilde{\nu} = \Delta\lambda/\lambda$, with $\Delta\lambda$ and $\Delta\tilde{\nu}$ being the spectral increments. NMR: Nuclear Magnetic Resonance; EPR: Electron Paramagnetic Resonance; VHF: Very High Frequency; UHF: Ultra High Frequency*

Instruments	*Platform*	*Launch Dates*	*Orbit details*	*Sight*	*References*
ACE-FTS	SCISAT-1	2003	Sun-synchronous, altitude 650 km, inclination 74°, repeat cycle 15 days	Solar occultation	http://www.ace.uwaterloo.ca/
AIRS	Aqua	2002	Sun-synchronous, altitude 705km, inclination 98.2 re-visit period twice a day	Nadir	http://airs.jpl.nasa.gov/
AVHRR	TIROS-N, NOAA 6, 7, 1214, 15, 16, 17 Metop	First 1978		Nadir	http://noaasis.noaa.gov/NOAASIS/ml/avhrr.html
CALIOP	CALIPSO	2006	2-wavelength at 532 and 1,064 nm	Nadir	https://calipso.cnes.fr/fr/CALIPSO/Fr/lidar.htm
CERES	AQUA	2002	Radiometer with 29 spectral bands from 0.4 to 14.4 μm	Nadir	http://ceres.larc.nasa.gov/aqua_ceres.php
CrIS	NPP	2011	Sun-synchronous, altitude 824km, inclination 98.7 re-visit period 101 min	Nadir	http://www.nasa.gov/mission_pages/NPP/mission_overview/index.html
HALOE	UARS	1991		Solar occultation	
IASI	Metop 1, 2 and 3	2006, 2012, 2016	Sun-synchronous, altitude 840 km, inclination 98.8 re-visit period twice a day	Nadir	http://smsc.cnes.fr/IASI/index.htm
IIR	CALIPSO	2006	3 spectral channels (8.7, 10.6 and 12 μm)	Nadir	https://calipso.cnes.fr/fr/CALIPSO/Fr/CP_04-2009.htm
IMG	ADEOS	1996	Sun-synchronous, altitude 830km, inclination 98.6 re-visit period 10 days	Nadir	http://www.eorc.jaxa.jp/en/hatoyama/satellite/sendata/img_e.html
MIPAS + SCIAMACHY	ENVISAT	2002	Sun-synchronous, altitude 782 km, inclination 98.5 re-visit period 6 days	Nadir + Limb + Solar occultation	https://earth.esa.int/instruments/mipas/ + https://earth.esa.int/instruments/sciamachy/
MODIS	Terra + Aqua	1999 and 2002	Sun-synchronous, altitude 705 km, inclination 98.2 re-visit period 1-2 days	Nadir	http://modis.gsfc.nasa.gov/
POLDER	PARASOL		9 spectral channels between 443 and 910 nm	Multi-angular	https://parasol.cnes.fr/fr/PARASOL/Fr/
SEVIRI	MSG	2005	Geostationary, altitude 36,000 km, scan repeat 15 min	Nadir	http://www.eumetsat.int/website/home/Satellites/CurrentSatellites/Meteosat/MeteosatDesign/index.html#sev
TANSO-FTS	GOSAT	01 Feb 2008	Sun-synchronous, altitude 666 km, inclination 98°, re-visit period 3 days	Nadir	http://www.gosat.nies.go.jp/index_e.html
TES + HIRDLS	Aura	2004	Sun-synchronous, altitude 705 km, inclination 98.2°, repeat cycle 16 days/re-visit period twice a day	Nadir + Limb	http://tes.jpl.nasa.gov/ + http://www.eos.ucar.edu/hirdls/

Table A.2a. *Some past and present spatial missions*

Instruments	*Platform*	*Launch Dates*	*Sight*	*References*
OCO-2	LeoStar-2	2014	Nadir	http://oco.jpl.nasa.gov/
IASI-NG	Metop-SG	2021	Nadir	http://smsc.cnes.fr/IASI-NG/index.htm
IRS	MTG	2018	Nadir	http://www.eumetsat.int/website/home/Satellites/FutureSatellites/MeteosatThirdGeneration/MTGDesign/index.html
3MI	Metop-SG	2021	Nadir	https://directory.eoportal.org/web/eoportal/satellite-missions/m/metop-sg

Table A.2b. *Examples of future missions*

Code	*Notes*	*Geometry*	*Spectrum*	*Output*	*Gaseous Absorption*	*RTE*	*Scatteing.*	*Polarization*	*Source*
LBLRTM	Sea surface emissivity	1D	IR to UV	Rad + Flx	LBL	DOM	yes	no	http://rtweb.aer.com/lblrtm.html
4A/OP	Fast code	1D	IR to NIR	Rad + Jac	LBL	DOM	yes	no	http://4aop.noveltis.com/
RFM	NLTE/ISO		IR to UV	Rad + Flx + Jac	LBL		no	no	http://eodg.atm.ox.ac.uk/RFM/
V-LIDORT	BRDF	1D	IR to UV	Rad + Jac	yes	DOM	yes	yes	http://www.rtslidort.com/mainprod_lidort.html
DISORT	Monochromatic	1D	IR to UV	Rad + Flx	yes	DOM	yes	yes	http://en.wikipedia.org/wiki/DISORT
RRTM	Rapid code	1D	IR to UV	Rad + Flx	CkD	DOM	yes		http://rtweb.aer.com/
STREAMER	BRDF	1D	IR to VIS	Rad + Flx	ESFT	DOM	yes		http://stratus.ssec.wisc.edu/streamer/
FLUXNET	Neural network	1D	VIS to IR	Flx	Band Model		yes	no	http://stratus.ssec.wisc.edu/fluxnet/
MODTRAN	Lunar illumination	1D	IR to UV	Rad + Flx	CkD	DOM	yes		http://www.MODTRAN5.com/
6SV	BRDF	1D	VIS to NIR	Rad	Band Model	SOS	yes	yes	http://6s.ltdri.org
MATISSE	3D scene simulation	1D/3D	IR to VIS	Rad + BT	CkDist		yes		http://matisse.onera.fr/accueil
RTTOV	Jacobians	1D	IR to MW	Rad + Flx + Jac	LUTs	Predictor	yes		http://nwpsaf.eu/deliverables/rtm/index.html

<table>
<tr><td>ARTS</td><td>nadir/limb</td><td>1D/2D/3D</td><td>MW</td><td>Rad + Jac</td><td>LBL or LUTs</td><td>MC + DOM</td><td></td><td>yes</td><td>http://www.sat.ltu.se/arts/</td></tr>
<tr><td>OSS</td><td>rapid code</td><td>1D</td><td>IR to MW</td><td>Rad + Flx + Jac</td><td>ESFT</td><td>LUTs</td><td></td><td></td><td>http://rtweb.aer.com/oss_frame.html</td></tr>
<tr><td>MOMO</td><td>Ocean/ atmosphere</td><td>1D</td><td>UV to IR</td><td>Rad + Flx</td><td>CkD</td><td>Matrix operator</td><td>yes</td><td>yes</td><td>Francois.Ravetta@latmos.ipsl.fr / juergen.fischer@fu-berlin.de</td></tr>
<tr><td>SHDOM</td><td>3D atmosphere</td><td>3D</td><td>IR toNIR</td><td>Rad + Flx</td><td>CkD</td><td>DOM + HS</td><td>yes</td><td>no</td><td>http://nit.colorado.edu/shdom.html</td></tr>
<tr><td>3DMCPOL</td><td>3D polarization</td><td>3D</td><td>VIS to IR</td><td>Rad</td><td>CkD</td><td>MC</td><td>yes</td><td>yes</td><td>Celine.Cornet@univ-lille1.fr</td></tr>
<tr><td>SkyGen 3D</td><td>3D cloud</td><td>3D</td><td>VIS to IR</td><td>Rad</td><td>CkD</td><td>SHDOM</td><td>yes</td><td></td><td>pierrick.bonafons@gmail.com / http://www.alyotech.com/</td></tr>
<tr><td>COART</td><td>Ocean/atmosphere</td><td>1D</td><td>VIS to NIR</td><td>Rad + Flx</td><td>CkD</td><td>DOM</td><td>yes</td><td>no</td><td>http://cloudsgate2.larc.nasa.gov/jin/rtnote.html</td></tr>
<tr><td>DART</td><td>3D landscapes</td><td>1D/2D/3D</td><td>VIS to IR</td><td>Flx</td><td>Band Model</td><td>DOM</td><td>yes</td><td></td><td>http://www.cesbio.upstlse.fr/dart/license/fr/getLicense.php</td></tr>
<tr><td>KARINE</td><td>net exchange rates/k-distributions</td><td>1D</td><td>IR</td><td>Flx</td><td>CkD</td><td>MC</td><td></td><td></td><td>http://www.meso-star.com/Produits.html/http://edstar.lmd.jussieu.fr/codes</td></tr>
<tr><td>MOSART</td><td>Transmission & radiance</td><td>1D</td><td>UV to MW</td><td>Rad + Trs</td><td>CkD</td><td>DOM</td><td>yes</td><td>no</td><td>http://www.cpi.com/products/mosart.html</td></tr>
<tr><td>SCIATRAN</td><td>BRDF, Raman Scattering, Ocean</td><td>1D</td><td>UV to NIR</td><td>Rad + Flx</td><td>LBL/band model</td><td></td><td>yes</td><td>yes</td><td>http://www.iup.uni-bremen.de/sciatran/</td></tr>
<tr><td>LIBRADTRAN</td><td colspan="2">Library for Radiative Transfer</td><td colspan="4" rowspan="3">Libraries dedicated to radiative transfer with a collection of data, codes and documents for calculation of solar and infrared radiation in the Earth's atmosphere</td><td colspan="3">http://www.libradtran.org/doku.php?id=start</td></tr>
<tr><td>ARTDECO</td><td colspan="2">Atmospheric Radiative Transfer Database for Earth Climate Observation</td><td colspan="3">http://loa-git.univ-lille1.fr/wikiloa_public/doku.php?id=artdeco:start</td></tr>
<tr><td>CRMT</td><td colspan="2">Community Radiative Transfer Model</td><td colspan="3">http://www.ssec.wisc.edu/~paulv/Fortran90/CRTM/Developmental/</td></tr>
</table>

Table A.3. *Non-exhaustive list of available Radiative Transfer Codes*

Name	*Type*	*Content*	*Source laboratry*	*Website*
	Surface emissivity	Global Infrared Land Surface Emissivity database from MODIS data	WISC	http://cimss.ssec.wisc.edu/iremis/
	Surface emissivity	collection ofemissivity measurementsof natural and manmade materials	UCSB	http://www.icess.ucsb.edu/modis/EMIS/html/em.html
ASTER spectral Library	Surface emissivity	compilation of over 2400 spectra of natural and manmade materials for the ASTER instrument	JPL	http://speclib.jpl.nasa.gov/
ADAM	Surface reflectances	A Surface Reflectance Database for ESA's Earth Observation Missions	ESA	http://adam.noveltis.com/
FDS: Fonds De Sol	Surface reflectances	Construction of a Global Database of Surface Reflectance and Emissivity at a scale of 500 m	LOA	http://loafds.univ-lille1.fr/fondsdesol/index_uk.html
POSTEL: Pole d'Observationdes Surfaces continentales par Teledetection	Reflectance and BRDF	Database of surface properties including reflectance derived from satellite data and BRDF from POLDER satellite data		http://postel.obs-mip.fr/?-land-cover-68-
OPAC: Optical Properties of Aerosols and Clouds	Aerosols and clouds	Database of aerosol and cloud optical properties	Ether	http://ether.ipsl.jussieu.fr/etherTypo/?id=989&L=0
SCATTPORT	Light Scattering Software	Light Scattering Information Portal for the light scattering community	DFG	http://www.scattport.org/index.php/light-scattering-software
	Codes for Cloud properties	Numerical algorithm (T-matrix) for scattering properties calculation of non-spherical particles	GISS	http://www.giss.nasa.gov/staff/mmishchenko/t_matrix.html
Ice Cloud Bulk Scattering Model	Cloud properties	Single-scattering properties of cirrus clouds derived from radiative transfer models and *in situ* measurements	WISC	http://www.ssec.wisc.edu/ice_models/
	Cloud properties	Single-scattering properties of cirrus clouds derived from airborne *in situ* measurements	Met Office	http://www.metoffice.gov.uk/research/people/anthony-j-baran
US Standard Atmospheres	Atmospheric Profiles	Standard Profiles for various latitude (T, P, H_2O and O_3)	AFGL/NASA	http://en.wikipedia.org/wiki/U.S._Standard_Atmosphere
TIGR: Thermodynamic Initial Guess Retrieval	Atmospheric Profiles	Dataset with 2,311 profiles (T, P, H_2O and O_3) of over 80,000 radiosondes	LMD	http://ara.abct.lmd.polytechnique.fr/
–	Atmospheric Profiles	Profiles from meteorological data available for various sites at global scale since 1973	UW	http://weather.uwyo.edu/upperair/sounding.html
LIBRADTRAN: LIBrary for Radiative transfer		Libraries dedicated to radiative transfer with a collection of data, code and document for calculation of solar and thermal radiation in the Earth's atmosphere	LMU	http://www.libradtran.org/doku.php?id=start
ARTDECO: Atmospheric Radiative Transfer Database for Earth Climate Observation			LOA	http://loa-git.univ-lille1.fr/wikiloa_public/doku.php?id=artdeco:start

Table A.4. *Digital resources for Radiative Transfer*

Bibliography

[AIR 01] AIRES F.C., PRIGENT W.B., ROSSOW *et al.*, "A new neural network approach including first-guess for retrieval of atmospheric water vapor, cloud liquid water path, surface temperature and emissivities over land from satellite microwave observations", *Journal of Geophysical Research*, vol. 106, pp. 14887–14907, 2001.

[AMB 36] AMBARTZUMIAN V., "The effect of the absorption lines on the radiative equilibrium of the outer layers of the stars", *Publications of the astronomical Observatory of University of Leningrad*, vol. 6, pp. 7–18, 1936.

[AND 01] ANDREAE M.O., MERLET P., "Emission of trace gases and aerosols from biomass burning", *Global Biogeochemical, Cycles*, vol. 15, no. 4, pp. 955–966, 2001.

[BEC 03] BECHTEL C., ZAHN A., "The isotopologue composition of water vapour: a powerful tool to study transport and chemistry of middle atmospheric water vapour", *Atmospheric Chemistry and Physics discussion*, vol. 3, pp. 3991–4036, 2003.

[BER 05] BERNATH P.F. *et al.*, "Atmospheric chemistry experiment (ace): mission overview", *Geophysical Research Letters,* vol. 32, pp. 1–5, 2005.

[BOI 13] BOICHU M., MENUT L., KHVOROSTYANOV D. *et al.*, "Inverting for volcanic SO2 flux at high temporal resolution using spaceborne plume imagery and chemistry-transport modelling: the 2010 Eyjafjallajökull eruption case study", *Atmospheric Chemistry and Physics*, vol. 13, pp. 8569–8584, 2013.

[BOI 15] BOICHU M.L., CLARISSE L., PÉRÉ J.-C. *et al.*, "Temporal variations of flux and altitude of sulfur dioxide emissions during volcanic eruptions: implications for long-range dispersal of volcanic clouds", *Atmospheric Chemistry and Physics*, vol. 15, pp. 8381–8400, 2015.

[BON 11] BONADONNA C., GENCO R., GOUHIER M. *et al.*, "Tephra sedimentation during the 2010 Eyjafjallajökull eruption (Iceland) from deposit, radar, and satellite observations", *Journal of Geophysical Research*, vol. 116, p. B12202, 2011.

[BOR 99] BORN M., WOLF E., *Principles of Optics*, 7th ed., Cambridge University Press, 1999.

[BOU 15] BOUCHER O., *Atmospheric Aerosols – Properties and Climate Impacts*, Springer, Netherlands, 2015.

[BOW 03] BOWEN G.J., REVENAUGH J., "Interpolating the isotopologic composition of modern meteoric precipitation", *Water Resources Res*earch, vol. 39, p. 1299, 2003.

[BRÉ 96] BRÉON F.M., BOUFFIÈS S., "Land surface pressure estimates from measurements in the oxygen A absorption band", *Journal of Applied Meteorology*, vol. 35, pp. 69–77, 1996.

[CLA 10a] CLARISSE L., HURTMANS D., PRATA A.J. *et al.*, "Retrieving radius, concentration, optical depth, and mass of different types of aerosols from high-resolution infrared nadir spectra", *Applied Optics*, vol. 49, no. 19, pp. 3713–3722, July 2010.

[CLA 10b] CLARISSE L., PRATA F., LACOUR J.-L. *et al.*, "A correlation method for volcanic ash detection using hyperspectral infrared measurements", *Geophysical Research Letters*, vol. 37, p. L19806, 2010.

[CLE 07] CLERBAUX C., HADJI-LAZARO J., TURQUETY S. *et al.*, "The IASI/MetOp mission: first observations and highlight of its potential contribution to the GMES Earth observation component", *Space Research Today*, vol. 168, pp. 19–24, 2007.

[CLO 89] CLOUGH S.A., KNEIZYS F.X., DAVIES R.W., "Line shape and the water vapor continuum", *Atmos. Res.*, vol. 23, pp. 229–241, 1989.

[COH 07] COHEUR P.-F., HERBIN H., CLERBAUX C. *et al.*, "Observation of a young biomass burning plume by the ACE-FTS: first reported measurements of C_2H_4, C_3H_6O, H_2CO and PAN by infrared occultation from space", *Atmospheric Chemistry and Physics*, vol. 7, pp. 7907–7932, 2007.

[COM 13] COMPIÈGNE M., C-LABONNOTE L., DUBUISSON P., "The phase matrix truncation impact on polarized radiance", *AIP Conference Proceedings*, vol. 1531, no. 1, p. 95, May 2013.

[DEH 87] DE HAAN J.F., BOSMA P.B., HOVENIER J.W., "The adding method for multiple scattering calculations of polarized light", *Astronomy and Astrophysics*, vol. 183, pp. 371–391, 1987.

[DEI 69] DEIRMENDJIAN D., *Electromagnetic Scattering on Spherical Polydispersions*, Elsevier, New York, pp. 75–119, 1969.

[DOP 14] DOPPLER L., PREUSKER R., BENNARTZ R. *et al.*, "k-bin and k-IR: k-distribution methods without correlation approximation for non-fixed instrument response function and extension to the thermal infrared – applications to satellite remote sensing", *Journal of Quantitative Spectroscopy and Radiative Transfer*, vol. 133, pp. 382–395, 2014.

[DUB 05] DUBUISSON P., GIRAUD V., CHOMETTE O. *et al.*, "Fast radiative transfer modeling for infrared imaging radiometry", *Journal of Quantitative Spectroscopy and Radiative Transfer,* vol. 95, no. 2, pp. 201–220, 2005.

[DUB 08] DUBUISSON P., GIRAUD V., PELON J. *et al.*, "Sensitivity of thermal infrared radiation at the top of the atmosphere and the surface to ice cloud microphysics", *Journal of Applied Meteorology and Climatology*, vol. 47, pp. 2545–2560, 2008.

[DUB 09] DUBUISSON P., FROUIN R., DESSAILLY D. *et al.*, "Estimating the altitude of aerosol plume over ocean from reflectance ratio measurements in the O_2 A-band", *Remote Sensing Environment*, vol. 113, pp. 1899–1911, 2009.

[DUB 14] DUBUISSON P., HERBIN H., MINVIELLE F. *et al.*, "Remote sensing of volcanic ash plumes from thermal infrared: case study analysis from SEVIRI, MODIS and IASI instruments", *Atmospheric measurement Techniques*, vol. 7, pp. 359–371, 2014.

[DUF 02] DUFRESNE J.L., RICCHIAZZI P., FOUQUART Y. *et al.*, "Longwave scattering effects of mineral aerosols", *J. Atmos. Sci.*, vol. 59, pp. 1959–1966, 2002.

[EDW 06] EDWARDS D.P., EMMONS L.K., GILLE J.C. *et al.*, "Satellite-observed pollution from Southern Hemisphere biomass burning", *J. Geophys. Res.*, vol. 111, p. D14312, 2006.

[ELL 91] ELLINGSON R.G., FOUQUART Y., "The intercomparison of radiation codes in climate models: an overview", *J. Geophys. Res.*, vol. 96, pp. 8925–8927, 1991.

[FIS 91] FISCHER J., GRASSL H., "Detection of cloud-top height from backscattered radiances within the oxygen A-band. Part. I: theoretical study", *Journal of Applied Meteorology*, vol. 10, pp. 1245–1259, 1991.

[FRA 00] FRASER R., LI Z., CIHLAR J., "Hotspot and NDVI differencing synergy (HANDS): a new technique for burned area mapping over boreal forest", *Remote Sensing of Environment*, vol. 74, pp. 362–376, 2000.

[FRO 06] FROMM M., TUPPER A., ROSENFELD D. *et al.*, "Violent pyro-convective storm devastates Australia's capital and pollutes the stratosphere", *Geophys. Res. Lett.*, vol. 33, p. L05815, 2006.

[GAR 12] GARNIER A., PELON J., DUBUISSON P. *et al.*, "Retrieval of cloud properties using CALIPSO imaging infrared radiometer. Part I: effective emissivity and optical depth", *J. Applied Meteor. and Climatol.*, vol. 51, no. 7, pp. 1407–1425, 2012.

[GAR 13] GARNIER A., PELON J., DUBUISSON P. *et al.*, "Retrieval of cloud properties using CALIPSO imaging infrared radiometer. Part II: effective diameter and ice water path", *J. Applied Meteor. and Climatol.*, vol. 52, no. 11, pp. 2582–2599, 2013.

[GIG 99] GIGLIO L., KENDALL J.D., JUSTICE C.O., "Evaluation of global fire detection algorithms using simulated AVHRR infrared data", *International Journal of Remote Sensing*, vol. 20, pp. 1947–1985, 1999.

[GOR 94] GORDLEY L.L., MARSHALL B.T., "LINEPAK: algorithm for modeling spectral transmittance and radiance", *J. Quant. Spectrosc. Radiat. Transfer*, vol. 52, no. 5, pp. 563–580, 1994.

[HAR 13] HARTMANN D.L. *et al.*, "Observations: atmosphere and surface", in STOCKER T.F., QIN D., PLATTNER G.-K. *et al. (eds)*, *Climate Change 2013: the Physical Science Basis. Contribution of Working Group I to the Fifth Assessment Report of the Intergovernmental Panel on Climate Change*, Cambridge University Press, Cambridge, UK and New York, NY, pp. 159–254, 2013.

[HEI 03] HEINTZENBERG J., RAES F., SCHWARTZ S., "Tropospheric aerosols", in BRASSEUR G., PRINN R.G., PSZENNY A.A.P. (eds), *The Changing Atmosphere: an Integration and Synthesis of a Decade of Tropospheric Chemistry Research*, Global Change – The Igbp Series, Springer, 2003.

[HER 66] HERZBERG G., *Electronic Spectra of Polyatomic Molecules*, Van Nostrand, New York, 1966.

[HER 07] HERBIN H.D., HURTMANS C., WESPES S. *et al.*, "Global distributions of stable water vapour isotopologues", *Atmospheric Chemistry and Physics*, vol. 7, pp. 3957–3968, 2007.

[HER 09] HERBIN H.D., HURTMANS C., CLERBAUX *et al.*, "$H_2{}^{16}O$ and HDO measurements with IASI/MetOp", *Atmospheric Chemistry and Physics*, vol. 9, pp. 9433–9447, 2009.

[HER 13] HERBIN H., C.-LABONNOTE L., DUBUISSON P., "Multispectral information from TANSO-FTS instrument – Part 2: applications to aerosol effect on greenhouse gas retrievals", *Atmos. Meas. Tech.*, vol. 6, pp. 3313–3323, 2013.

[HOB 03] HOBBS P.V., SINHA P., YOKELSON R.J. *et al.*, "Evolution of gases and particles from a savanna fire in South Africa", *J. Geophys. Res.*, vol. 108, p. 8487, 2003.

[HOR 89] HORNIK K., STINCHCOMBE M., WHITE H., "Multilayer feed forward networks are universal approximators", *Journal Neural Network*, vol. 2, pp. 359–366, 1989.

[HUN 71] HUNT G.E., "A review of computational techniques for analysing the transfer of radiation through a model cloudy atmosphere", *J. Quant. Spectrosc. Radiat. Transfer*, vol. 11, no. 6, pp. 655–690, 1971.

[ICA 01] INTERNATIONAL CIVIL AVIATION ORGANIZATION (ICAO), Manual on volcanic ash, radioactive material and toxic chemical clouds, ICAO Doc. 9766-AN/954, Montreal, Quebec, Canada, 2001.

[INO 85] INOUE T., "On the temperature and effective emissivity determination of semi-transparent cirrus clouds by bi-spectral measurements in the window region", *J. Meteorol. Soc. Jpn.*, vol. 63, pp. 88–98, 1985.

[INO 87] INOUE T., "A cloud type classification with NOAA 7 split-window measurements", *J. Geophys. Res.*, vol. 92, pp. 3991–4000, 1987.

[IPC 13] IPCC, "Summary for policymakers", in STOCKER T.F., QIN D., PLATTNER G.-K. *et al.* (eds), *Climate Change 2013: the Physical Science Basis. Contribution of Working Group I to the Fifth Assessment Report of the Intergovernmental Panel on Climate Change*, Cambridge University Press, Cambridge, UK and New York, p. 1535, 2013.

[JAC 11] JACQUINET-HUSSON N., CREPEAU L., ARMANTE R. *et al.*, "The 2009 edition of the GEISA spectroscopic database", *J. Quant. Spectrosc. Radiat. Transf.*, vol. 112, pp. 2395–2445, 2011.

[JOU 97] JOUZEL J., ALLEY R.B., CUFFEY K.M. *et al.*, "Validity of the temperature reconstruction from water isotopologues in ice cores", *J. Geophys. Res.*, vol. 102, p. 26471, 1997.

[KEI 08] KEIM C., LIU G.Y., BLOM C.E. *et al.*, "Vertical profile of peroxyacetyl nitrate (PAN) from MIPAS-STR measurements over Brazil in February 2005 and its contribution to tropical UT NOy partitioning", *Atmos. Chem. Phys.*, vol. 8, pp. 4891–4902, 2008.

[KOK 10] KOKHANOVSKY A.A., BUDAK V.P., CORNET C. *et al.*, "Benchmark results in vector atmospheric radiative transfer", *J. Quant. Spectrosc. Radiat. Transf.*, vol. 111, pp. 1931–1946, 2010.

[KUA 03] KUANG Z.M., TOON G.C., WENNBERG P.O. *et al.*, "Measured HDO/H_2O ratios across the tropical tropopause", *Geophys. Res. Lett.*, vol. 30, p. 1372, 2003.

[KUR 95] KURUCZ R.L., "The solar irradiance by computation", in ANDERSON G.P. *et al.* (eds), *Proceedings of the 17th Annual Conference on Atmospheric Transmission Models*, Phillips Laboratory, Hanscom Air Force Base, PL-TR-95–2060, pp. 333–334, 1995.

[LAC 91] LACIS A.A., OINAS V., "A description of the correlated k distributed method for modelling nongray gaseous absorption, thermal emission, and multiple scattering in vertically inhomogeneous atmospheres", *J. Geophys. Res.*, vol. 96, pp. 9027–9063, 1991.

[LEB 39] LEBEDINSKY A.I., "Radiative equilibrium in the Earth's atmosphere", *Proceedings of the Leningrad University Series Mathematics*, vol. 3, no. 31, pp. 152–175, 1939.

[LEF 86] LEFEBVRE-BRION H., FIELD R.W., *Perturbations in the Spectra of Diatomic Molecules*, Academic Press, 1986.

[LEN 07] LENOBLE J., HERMAN M., DEUZÉ J.-L. *et al.*, "A successive order of scattering code for solving the vector equation of transfer in the earth's atmosphere with aerosols", *J. Quant. Spectrosc. Radiat. Transf.*, vol. 107, no. 3, pp. 479–507, 2007.

[LET 07] LE TREUT H., SOMERVILLE R., CUBASCH U. *et al.*, "Historical overview of climate change", in SOLOMON S., QIN D., MANNING M. *et al.* (eds), *Climate Change 2007: the Physical Science Basis, Contribution of Working Group I to the Fourth Assessment Report of the Intergovernmental Panel on Climate Change*, Cambridge University Press, Cambridge, UK and New York, NY, 2007.

[MAR 05] MARSHAK A., DAVIS A.B. (eds), *3D Radiative Transfer in Cloudy Atmosphere*, Springer, New York, 2005.

[MAS 06] MASON S.A., TRENTMANN J., WINTERRATH T. *et al.*, "Intercomparison of two box models of the chemical evolution in biomass-burning smoke plumes", *J. Atmos. Chem.*, vol. 55, pp. 273–297, 2006.

[MAU 98] MAUZERALL D.L., LOGAN J.A., JACOB D.J. *et al.*, "Photochemistry in biomass burning plumes and implications for tropospheric ozone over the tropical South Atlantic", *J. Geophys. Res.*, vol. 103, no. 19, pp. 8401–8424, 1998.

[MCC 72] MCCLATCHEY R.A., FENN R.W., SELBY J.E.A. *et al.*, *Optical Properties of the Atmosphere*, 3rd ed., AFCRL 72–0497, AD 753075, 1972.

[MIS 96] MISHCHENKO M.I., TRAVIS L.D., MACKOWSKI D.W., "T-matrix computations of light scattering by nonspherical particles: a review", *J. Quant. Spectrosc. Radiat. Transfer*, vol. 55, pp. 535–575, 1996.

[MIS 00] MISHCHENKO M.I., "Calculation of the amplitude matrix for a nonspherical particle in a fixed orientation", *Appl. Opt.*, vol. 39, pp. 1026–1031, 2000.

[MOY 96] MOYER E.J., IRION R.W., YUNG Y.L. *et al.*, "ATMOS stratospheric deuterated water vapour and implications for troposphere-stratosphere transport", *Geophys. Res. Lett.*, vol. 23, p. 2385, 1996.

[MÜL 03] MÜLLER M.D., KAIFEL A.K., WEBER M. *et al.*, "Ozone profile retrieval from global ozone monitoring experiment (GOME) data using a neural network approach (Neural Network Ozone Retrieval SYstem (NNORSY))", *J. Geophys. Res.*, vol. 108, 2003.

[NEC 84] NECKEL H., LABS D., "The solar spectrum between 3300 and 12500 nm", *Solar Physics*, vol. 90, pp. 205–258, 1984.

[PAR 91] PAROL F., BURIEZ J.C., BROGNIEZ G. *et al.*, "Information content of AVHRR channels 4 and 5 with respect to the effective radius of cirrus cloud particles", *J. Appl. Meteorol.*, vol. 30, pp. 973–984, 1991.

[PEY 10] PEYRIDIEU S., CHÉDIN A., TANRÉ D. *et al.*, "Saharan dust infrared optical depth and altitude retrieved from AIRS: a focus over North Atlantic – comparison to MODIS and CALIPSO", *Atmos. Chem. Phys.*, vol. 10, pp. 1953–1967, 2010.

[PRA 89] PRATA A.J., "Infrared radiative transfer calculations for volcanic ash clouds", *Geophys. Res. Lett.*, vol. 16, pp. 1293–1296, 1989.

[PRA 07] PRATA A.J., BERNARDO C., "Retrieval of volcanic SO_2 column abundance from atmospheric infrared sounder data", *J. Geophys. Res.: Atmospheres*, vol. 112, 2007.

[PRA 09] PRATA A.J., TUPPER A.T., "Aviation hazards from volcanoes: the state of the science", *Natural Hazards*, vol. 51, pp. 239–244, 2009.

[PRA 12] PRATA F.J., PRATA A.T., "Eyjafjallajokull volcanic ash concentrations determined using SEVIRI measurements", *J. Geophys. Res.*, vol. 117, p. D00U23, 2012.

[PUJ 12] PUJOL O., BROGNIEZ G., LABONNOTE L., "Simplification for Fraunhofer diffracting pattern of various randomly oriented ice crystals in cirrus", *Journal of the Optical Society of America A*, vol. 29, no. 9, pp. 2015–2018, 2012.

[PUJ 15] PUJOL O., "Comment on the (misused) concept of photon in radiative transfer, and proposition of a neologism", *J. Quant. Spectrosc. Radiat. Transfer*, vol. 159, pp. 29–31, 2015.

[RIE 10] RIEDI J., MARCHANT B., PLATNICK S. *et al.*, "Cloud thermodynamic phase inferred from merged POLDER and MODIS data", *Atmos. Chem. Phys.*, vol. 10, pp. 11851–11865, 2010.

[ROD 00] RODGERS C.D., *Inverse Methods for Atmospheric Sounding: Theory and Practice*, World Sci., Hackensack, NJ, 2000.

[ROS 58] ROSENBLATT F., "The perceptron: a probabilistic model for information storage and organization in the brain", *Psychological Review*, vol. 65, pp. 386–408, 1958.

[ROS 91] ROSSOW W.B., SCHIFFER R.A., "ISCCP cloud data products", *Bulletin American Meteorology Society*, vol. 71, pp. 2–20, 1991.

[ROT 13] ROTHMAN L.S., GORDON I.E., BABIKOV Y. *et al.*, "The HITRAN 2012 molecular spectroscopic database", *J. Quant. Spectrosc. Radiat. Transfer*, vol. 130, pp. 4–50, 2013.

[SAU 10] SAUNDERS R., GEER A., RAYER P. *et al.*, RTTOV-9 science and validation report, NWPSAF-MO-TV-020, 2010.

[SIL 96] SILVESTER P.P., FERRARI R.L., *Finite Elements for Electrical Engineers*, 3rd ed., Cambridge University Press, 1996.

[STA 88] STAMNES K., TSAY S.C., WISCOMBE W. *et al.*, "Numerically stable algorithm for discrete-ordinate-method radiative transfer in multiple scattering and emitting layered media", *Appl. Opt.*, vol. 27, no. 12, pp. 2502–2509, 1988.

[THO 02] THOMAS G.E., STAMNES K., *Radiative Transfer in the Atmosphere and Ocean*, Cambridge University Press, 2002.

[THO 10] THOMAS H.E., WATSON I.M., "Observations of volcanic emissions from space: current and future perspectives", *Natural Hazards*, vol. 54, pp. 323–354, 2010.

[THU 03] THUILLIER G., HERSÉ M., SIMON P.C. *et al.*, "The solar spectral irradiance from 200 to 2400 nm as measured by the SOLSPEC spectrometer from the ATLAS 1–2–3 and EURECA missions", *Solar Physics*, vol. 214, no. 1, pp. 1–22, 2003.

[ULB 83] ULBRICH C.W., "Natural variations in the analytical form of the raindrop size distribution", *Journal of Climate Applied Meteorology*, vol. 22, pp. 1764–1775, 1983.

[VAN 81] VAN DE HULST H.C., *Light Scattering by Small Particles*, Dover, 1981.

[VAN 14] VAN DAMME M., CLARISSE L., HEALD C.L. *et al.*, "Global distributions, time series and error characterization of atmospheric ammonia (NH_3) from IASI satellite observations", *Atmos. Chem. Phys.*, vol. 14, pp. 2905–2922, 2014.

[WEB 03] WEBSTER C.R., HEYMSFIELD A.J., "Water isotopologue ratios D/H, 18O/16O, 17O/16O in and out of clouds map dehydration pathways", *Science*, vol. 302, pp. 1742–1745, 2003.

[WEN 94] WEN S., ROSE W.I., "Retrieval of sizes and total masses of particles in volcanic clouds using AVHRR channels 4 and 5", *J. Geophys. Res.*, vol. 99, pp. 5421–5431, 1994.

[WEN 12] WENDISCH M., YANG P., *Theory of Atmospheric Radiative Transfer*, 1st ed., Wiley-VCH, 16 April 2012.

[WOR 06] WORDEN J., BOWMAN K., NOONE D. *et al.*, "Tropospheric emission spectrometer observations of the tropospheric HDO/H_2O ratio: estimation approach and characterization", *J. Geophys. Res.*, vol. 111, 2006.

[YAN 98] YANG P., LIOU K.N., "Single-scattering properties of complex ice crystals in terrestrial atmosphere", *Contributions Atmospheric Phys*ics, vol. 71, pp. 223–248, 1998.

[YAN 05] YANG P., WEI H., HUANG H.-L. *et al.*, "Scattering and absorption property database for nonspherical ice particles in the near-through far-infrared spectral region", *Appl. Opt.*, vol. 44, pp. 5512–5523, 2005.

[YOK 03] YOKELSON R.J., BERTSCHI I.T., CHRISTIAN T.J. *et al.*, "Trace gas measurements in nascent, aged, and cloud-processed smoke from African savanna fires by airborne Fourier transform infrared spectroscopy (AFTIR)", *J. Geophys. Res.*, vol. 108, p. 8478, 2003.

[YUE 09] YUE Q., LIOU K.N., "Cirrus cloud optical and microphysical properties determined from AIRS infrared spectra", *Geophys. Res. Lett.*, vol. 36, 2009.

[ZAK 04] ZAKHAROV V.I., IMASU R., GRIBANOV K.G. *et al.*, "Latitudinal distribution of the deuterium to hydrogen ratio in the atmospheric water vapour retrieved from IMG/ADEOS data", *Geophys. Res. Lett.*, vol. 31, p. L12104, 2004.

[ZEH 10] ZEHNER C., "Monitoring volcanic ash from space", *Proceedings of the ESA-EUMETSAT Workshop on the 14 April to 23 May 2010 Eruption at the Eyjafjoll Volcano*, ESA-Publication STM-280, South Iceland, Frascati, Italy, 26–27 May 2010.

Index

G, I, L

P, R

S, V

Other titles from

iSTE

in

Earth Systems – Environmental Engineering

2015

MERCIER-LAURENT Eunika
The Innovation Biosphere

MONACO André, PROUZET Patrick
Seas and Oceans Set
Volume 7 – Marine Ecosystems: Diversity and Functions

2014

DE LARMINAT Philippe
Climate Change: Identification and Projections

MONACO André, PROUZET Patrick
Seas and Oceans Set
Volume 1 – Ocean in the Earth System
Volume 2 – The Land-Sea Interactions
Volume 3 – Ecosystem Sustainability and Global Change
Volume 4 – Vulnerability of Coastal Ecosystems and Adaptation
Volume 5 – Development of Marine Resources
Volume 6 – Value and Economy of Marine Resources

MUTTIN Frédéric
Marine Coastal and Water Pollutions: Oil Spill Studies

2013

LÉCUYER Christophe
Water on Earth: Physicochemical and Biological Properties

LEGCHENKO Anatoly
Magnetic Resonance Imaging for Groundwater

2012

LE MENN Marc
Instrumentation and Metrology in Oceanography

PAVÉ Alain
Modeling of Living Systems: From Cell to Ecosystem

2010

TANGUY Jean-Michel
Physical Processess and Measurement Devices – EHS volume 1

TANGUY Jean-Michel
Mathematical Models – EHS volume 2

TANGUY Jean-Michel
Numerical Methods – EHS volume 3

TANGUY Jean-Michel
Practical Applications in Engineering – EHS volume 4

TANGUY Jean-Michel
Modeling Software – EHS volume 5

Printed and bound by CPI Group (UK) Ltd, Croydon, CR0 4YY

13/09/2026

14961077-0002